BASIC ANTHROPOLOGY UNITS

GENERAL EDITORS
George and Louise Spindler
STANFORD UNIVERSITY

PRIMATE BEHAVIOR AND THE EMERGENCE OF HUMAN CULTURE

JANE BECKMAN LANCASTER
Delta Regional Primate Research Center

Primate Behavior and the Emergence of Human Culture

HOLT, RINEHART AND WINSTON
New York   Chicago   San Francisco   Atlanta
Dallas   Montreal   Toronto   London   Sydney

COPYRIGHT ACKNOWLEDGMENTS

Figures reproduced courtesy of the following:

1–1a, b; 1–2c   From S. L. Washburn and Ruth Moore, *Ape into Man: A Study of Human Evolution*, pp. 26, 27, 67. Copyright © 1974 by Little, Brown and Company (Inc.). Reprinted by permission.

1–2a, b   From the *American Journal of Physical Anthropology*, 26:119–130. Reprinted by permission of The Wistar Press and Milton Hildebrand.

2–1; 2–2   From "Baboon social behavior" by K. R. L. Hall and Irven DeVore, in *Primate Behavior: Field Studies of Monkeys and Apes* edited by Irven DeVore. Copyright © 1965 by Holt, Rinehart and Winston, Inc. Reprinted by permission of Irven DeVore, Anthro-Photo File, and Holt, Rinehart and Winston, Inc.

2–3   From L. A. Rosenblum, I. C. Kaufman, and A. J. Stynes, *Animal Behavior*, 12:338–342, by permission of Baillière Tindall, publishers, and I. Charles Kaufman.

2–4   From I. C. Kaufman and L. A. Rosenblum, "Effects of separation from mother on the emotional behavior of infant monkeys," *Annals New York Academy of Sciences*, 159:681–695. Reprinted by permission of The New York Academy of Sciences and I. Charles Kaufman.

2–5   From *Primates*, 1970, 11:335–390. Reprinted by permission of the publisher and Naoki Koyama.

2–6   Courtesy of Geza Teleki.

2–7   From A. Schultz, 1969, *The Life of Primates*, p. 149, by permission of Weidenfeld & Nicolson and the author.

3–1   Courtesy of A. Suzuki.

4–1   From "Communication in monkeys and apes" by Peter Marler, in *Primate Behavior: Field Studies of Monkeys and Apes*, edited by Irven DeVore. Copyright © 1965 by Holt, Rinehart and Winston, Inc. Reprinted by permission of the author and Holt, Rinehart and Winston, Inc.

4–2   From "Language and the brain" by Norman Geschwind. Copyright © April 1972 by Scientific American, Inc. All rights reserved.

5–1   From J. van Lawick-Goodall, *In the Shadow of Man*, by permission of Houghton Mifflin Company, William Collins Sons and Company, London, and Baron Hugo von Lawick. Photograph by Geza Teleki.

Library of Congress Cataloging in Publication Data

Lancaster, Jane Beckman, 1935–
Primate behavior and the emergence of human culture.

(Basic anthropology units)
Bibliography: p. 91
Includes index.
1. Primates—Behavior. 2. Human evolution.
3. Culture—Origin. 4. Social behavior in animals.
5. Psychology, Comparative. I. Title.
GN280.7.L36   573.2   74-22256
ISBN 0–03–091311–X

To Sherry Washburn, teacher and friend

Foreword

THE BASIC ANTHROPOLOGY UNITS

Basic Anthropology Units are designed to introduce students to essential topics in the contemporary study of man. In combination they have greater depth and scope than any single textbook. They may also be assigned selectively to cover topics relevant to the particular profile of a given course, or they may be utilized separately as authoritative guides to significant aspects of anthropology.

Many of the Basic Anthropology Units serve as the point of intellectual departure from which to draw on the Case Studies in Cultural Anthropology and Case Studies in Education and Culture. This integration is designed to enable instructors to utilize these easily available materials for their instructional purposes. The combination introduces flexibility and innovation in teaching and avoids the constraints imposed by the encyclopedic textbook. To this end, selected Case Studies have been annotated in each unit. Questions and exercises have also been provided as suggestive leads for instructors and students toward productive engagements with ideas and data in other sources as well as the Case Studies.

This series was planned over a period of several years by a number of anthropologists, some of whom are authors of the separate Basic Units. The completed series will include units representing all the basic sectors of contemporary anthropology, including archeology, biological anthropology, and linguistics, as well as the various subfields of social and cultural anthropology.

THE AUTHOR

Jane Beckman Lancaster is currently affiliated with the Delta Regional Primate Research Center, Covington, Louisiana, and lectures at the University of New Orleans. She did her major field work in the Zambezi Valley, Zambia, in 1967–1969, on the social behavior of the vervet monkey, *Cercopithecus aethiops*. She received her Ph.D. in anthropology from the University of California at Berkeley. Her writings include articles on the evolution of tool-using behavior and on primate communication systems and the emergence of human language, both subjects of great importance to the present Unit. She is married to an anthropologist and they have two sons.

THIS UNIT

When we first started teaching Introductory Anthropology in the early 1950s at Stanford we spent some time on human evolution but very little on primate life and its relationship to the emergence of human culture. Now when we teach the course we spend the first two weeks of the quarter on this topic. Students seem to learn much from this exposure. Perhaps we are as species-centric as we are ethnocentric. The perspective that anthropology can claim as one of its major values seems particularly reinforced by a substantial consideration of primate behavior and its relationship to human behavior.

Primate studies began a meteoric rise to significance mostly after 1958. Since then at least 30 major studies have been done in the field and we know a great deal about the apes and the New and Old World monkeys, though some species are much better known than others. Information and understanding gained from these studies about primate communication, social organization, ecological adaptation, and psychology is complex and is constantly being modified by new findings. It is extremely difficult, therefore, to bring this material together in a meaningful integration, especially within brief compass. Jane Lancaster has done it well in this Basic Anthropology Unit.

Students reading this Unit will be convinced that man is not a primate by chance. Many of the features which we think of as being earmarks of human society are shared by the other primates, and they certainly existed long before humans appeared on the scene. The social behavior and ecological adaptations of monkeys and apes in the far distant past set the stage for the emergence of the human way of life. An understanding of these relationships helps the student to better understand human life. The author exploits her material for its relevance to this understanding as she moves through discussions of social bonding, sex behavior, tool use, signals, communication, and language, and brings it sharply into focus with her final chapter, which discusses some direct relationships to pressing contemporary social problems.

Jane Lancaster provides useful suggestions for topics to discuss and projects to do at the end of each chapter. She also includes an excellent list of recommended readings and a list of Case Studies in Cultural Anthropology that provide good ethnographic data on human behavior patterns in small groups, that can be used to extend to the analysis of human groups some of the understandings developed in this Unit. A list of useful films on primates is also included. Though this Unit is written in understandable, clear, and nontechnical language, a glossary including every word that is used in any special sense in this Unit is provided at the end of the text.

George and Louise Spindler
General Editors
PHLOX, WISCONSIN

Preface

The purpose of this book is to present an introduction to the understanding of man and human behavior as the products of a long evolutionary history. This perspective is one which belongs to no single field of science. Contributions have come from both the natural and social sciences, from biologists, zoologists, anatomists, anthropologists, psychologists, and sociologists to name only a few. However, the special insights of each of these sciences are unified by a single theme, the adaptive nature of human biology and behavior. We hope that this perspective will give us fresh insight into the human condition and the problems facing our species today.

In writing the book, I have drawn on materials from the rapidly expanding field of primatology, which has received fresh impetus from field studies of the natural behavior of primates in the wild during the past decade. I have not tried to summarize the present state of the field but rather to point out areas which are most relevant to understanding man and human behavior. There is clearly much more to be learned and understood; this book represents the barest beginnings of a field which is only starting to mature.

In writing about the fossil record of the earliest human beings, the Australopithecines, I have tried to present only the broadest outlines and not become involved in the details of interpretation of the fossils. There have been a number of very important fossils discovered in East Africa during the past few years and there is every indication that the earliest period of human history, from ten million to one million years ago, is being rapidly filled in. However, there is much more work and analysis to be done before final interpretations can be made, so this book discusses the general shape of the human adaptive pattern and the fossil record without going into the many controversies of interpretation surrounding virtually every new fossil.

The field research on which this book draws for illustration and insight was supported by a National Science Foundation grant (GS 1414) for the study of vervet monkey social behavior in Zambia, Africa, 1967–1969.

I particularly wish to thank Lily Arboleta and Mary Fisher for help in preparing the manuscript. I would also like to thank the following field workers and researchers for their generosity in letting me use their photographs for illustrations: Geza Teleki, Pennsylvania State University; Irven DeVore, Harvard University; I. Charles Kaufman, University of Colorado Medical Center; and Akira Suzuki, Kyoto University. I would also like to thank Thelma Rowell, the University of California, Berkeley, and C. S. Lancaster, Tulane University, for helpful discussions and suggestions on some of the ideas found in these chapters. Finally, my gratitude goes to David P. Boynton of Holt, Rinehart and Winston, Inc., and George and Louise Spindler, General Editors of the Basic Anthropology Unit Series, for their patience and helpful suggestions in the preparation of the manuscript.

J. B. L.

Contents

The Evolution of Behavior

PERSPECTIVE

The purpose of this book is to present a perspective on the human species and human behavior as the product of a long, complex evolutionary history. Mankind has evolved and expanded in accordance with the same major evolutionary processes as have other species of animal life. The human species' understanding of itself and its place in nature has too long been based on distinctions between itself and other forms of life: first, as a special, divine creation destined to dominate the earth, and more recently as the sole bearer of culture, a vast body of learned behavior supposedly freeing humanity from the demands of the environment and natural forces. Today, however, the mounting modern problems of international warfare and genocide, urban crowding, oppression of minorities, emotional isolation of the individual, and world energy and food shortages make it abundantly clear that our species is neither free of itself nor its environment. Like all other creatures mankind must conform to the evolutionary processes of natural selection and adaptation. The prospects of world overpopulation and the collapse of energy and food supplies suggest that in fact our species may be heading into a period of extreme environmental pressure and selection. Now more than ever we need to understand ourselves and our relationship to the rest of the natural world.

THE EVOLUTIONARY PROCESS

Evolution is a process by which those individuals who are most fit survive and pass on their genes to the next generation. There are many different ways to be "most fit," and it does not necessarily mean those who are the strongest and most aggressive. At any given time selective advantage may come from such various factors as special care and interest in the young infant, adventurousness or conservatism in the food quest, resistance to disease, or perhaps excellent digestion. It all depends on what particular factors in the environment of the group are most significant at the time in affecting the success of individuals in surviving into their reproductive years and in having and raising offspring.

Take, for example, skin color. When under stimulation from the sun's rays, the human skin manufactures a substance known as vitamin D which is essential to normal health and growth but which can be poisonous in too large amounts. Darkened skin, whether from tanning or from inheritance, provides protection from a vitamin D overdose. The darkness comes from a thin layer of melanin in the skin, which screens the sun's rays from reaching the deeper layers where vitamin D is synthesized. In parts of the world where the sun's rays are very strong, there has been selective pressure in favor of individuals with dark skin. This was particularly true in time past when tailored clothing, hats, and sun lotion were unknown. We can expect, then, that there will be a strong tendency for human groups who have lived near the equator for a long time to have dark skins. A world survey shows that this is exactly the case. Peoples living near the equator in Africa, India, the southern Pacific Ocean, and even in the New World all have darker skins than their neighbors and close relatives living in more temperate climates. The skin of recent migrants to the tropics such as Europeans or Asiatics has not been darkening because, when they moved to their new environments, they came after the development of tailored clothing. These groups will never evolve dark skin as long as they maintain behavior patterns which protect their skins from too much sun.

Populations in a given area are never uniform. For a variety of reasons there is always genetic variation within a population. Natural selection is an imperfect process, not an efficient sifter of each generation. Individuals can be successful for different reasons, one because he was aggressive and another because he was cautious at crucial junctures in life. If natural selection is consistent through the generations, there will be a gradual increase in gene frequencies of a particular adaptive trait until a large proportion of the population possesses it. However, there is nothing fixed or immutable in the high frequency of a particular gene. Its adaptive advantage might disappear tomorrow and the direction of natural selection be reversed. We need to emphasize the point that the genetic composition of a population at any one time represents successes of time past. Humanity or any other species is genetically prepared for the present only to the extent that it has not changed from the past.

Natural selection never stops affecting the genetic composition of each generation, but the way it affects it will change as circumstances change. Early in human history, before the species had evolved efficient upright walking, the big toe was opposable to the other toes to help in climbing, as it is today in monkeys and apes. As the human species gradually began to move into a way of life which called for efficient bipedalism, there must have been selective advantage for individuals whose big toes were less in opposition to the others so they could be used to bear more of the body's weight. These individuals could walk farther and run faster without tiring compared to others with more apelike feet. Eventually, the genetic elements which determined the apelike foot were selected out of the gene pool and the range of variation in foot form shifted from the wide foot with opposable big toe to the long, narrow foot of the modern human.

Today most people are born with feet whose form is "acceptable" for bipedal walking, and natural selection is affecting other aspects of human anatomy and behavior. For example, in 1918–1919 there was a worldwide epidemic of Spanish influenza killing 20 million people, many more deaths than from World War I. Every year new flu viruses, some highly virulent and others only bothersome, sweep around the world, passing from country to country until most human populations have been exposed. Clearly natural selection is still at work shaping and changing man. There is no such thing as a form of life reaching static, immutable perfection. As long as the environment of a species varies, and the relationships between species fluctuate and change, there will be changes in forms of life.

In order to understand the biology and behavior of human beings, we have got to know how and why we got to be the way we are today. We have to discover what successes and failures of time past were in order to understand the species we find in time present. An appreciation of the factors which shaped modern man may help us to better understand and find solutions for problems confronting him today.

RECONSTRUCTING THE PAST

There are really only two lines of evidence which can be used for reconstructing the evolutionary history of a species. The first is the fossil and archeological record itself, that is, any aspect of the past behavior and anatomy of a species which has been accidentally preserved instead of destroyed by the forces of nature. Such relics of the past can be the actual bony remains of an individual, or a footprint, a tool, the remains of a meal, or even the microscopic bits of pollen floating in the air at the time such things were first covered with earth. All of these relics aid in understanding what a species was like and how it lived. However, this record of the past is very imperfect because it is based on accidental preservation. The further back in time one goes, the less complete the record, because of the longer period of time which has elapsed, allowing the processes of weathering and movement in the earth's crust to destroy the fossils. Furthermore, many aspects of behavior do not leave any trace in the fossil record. Vital questions about the nature and evolution of human emotions cannot be answered from the fossil record alone. Even such basic questions as to diet and how food was obtained may not appear, or the indications may not be readily intelligible.

Fortunately, there are other lines of evidence which can aid in reconstructing the past. These leads come from what is called the "comparative method." The world is full of potential comparisons; the trick is to choose the one most apt to help answer the questions asked. In terms of better understanding humans, there are three major groupings of animals whose behavior can be studied to yield useful comparisons. The first is various contemporary groups of the human species. The world is home for many different kinds of human beings living in different ecological adaptations with very different cultural traditions. If we can show that a particular behavior pattern such as smiling is a universal sign of positive emotions, then we have discovered something important about the nature of mankind. Or we can study particular groups of humans living today in ways similar to the life of early man. The few surviving groups living a gathering-hunting* way of life may help us to learn something about the quality of social and emotional life of time past.

* The term "gathering-hunting" is used instead of the more usual reversed form "hunting-gathering" because of the strong tendency for students to overemphasize hunting in their minds as the keystone to the human adaptive pattern.

The second line of evidence comes from comparisons with our closest relatives, the Old World monkeys and apes. It has been an absolute minimum of 5 to 6 million years since we have shared a common ancestor with any of the living species of ape and for the monkeys it has been closer to 20 million or more years. Nevertheless, in terms of evolutionary history it has not been that long and many useful comparisons can be made. For example, the evolution of tool using and manufacture clearly played an important role in the evolution of man. Until a few years ago most people thought that this behavior was unique to humans, but now we know that the wild chimpanzee makes and uses tools in a wide variety of situations. Since the gorilla and the chimpanzee are our closest living relatives, it is useful to study the situations in which the chimpanzee uses tools, how he uses them, and why the gorilla does not. Answers to such questions can help us in reconstructing the adaptive situation which led the human species to commit itself to a tool using way of life.

There is one final group of animals which yield useful comparisons, the group-living hunters such as wild dogs and lions. In contrast to humans, monkeys and apes are basically committed to a diet of fruit, vegetables, nuts, and seeds, with an occasional bit of meat. Meat, however, is an important part of the human diet, and the archeological record suggests that meat eating and tool using may occur at the earliest stages of our species' separation from the apes. When considering such factors as the size of early man's home range, his food-sharing behavior, or his hunting techniques, we must turn to the other group-living hunters for relevant data.

We are not saying that humans are always like monkeys, apes, or carnivores. Rather we are saying that the right comparisons help us to understand what our species really is. They make us aware of major long-term continuities which link the behavior of humans to our nearest relatives, as well as helping to illuminate ways in which we are unique. For example, many emotional processes in humans, such as the way an infant learns to love and bond first to its mother and later its family, are similar in monkeys, apes, and humans. When we are concerned with questions about mental health and normal development in humans, we can turn to monkeys and apes for information or controlled experimentation, whereas data from rats or guinea pigs may have very little relevance to understanding the human situation.

We have just as much to learn from the nonhuman primates when we make comparisons which give negative answers, ones that say that humans are very different. If we find a behavioral characteristic which seems to set us off from other primates, then we can guess that we may be dealing with a behavior which is fundamental to the human pattern—perhaps one that was responsible for the original separation of humans from the apes. It was mentioned above that the diet and food-getting behavior of humans sets them off from the other primates. It sets them off from the group-living carnivores as well. Our species is neither vegetarian nor carnivorous; we are something unique—gatherer-hunters. Before modern times our diet depended on the collecting and hunting of both plant and animal foods shared between members of the same social group. We can only really appreciate this fact when we can see how different it is from the life style of other animals.

There is one final type of comparison which yields answers which point up neither similarities nor differences between humans and other animals. These fall in a category we shall call "anachronisms." Anachronisms occur because of discontinuities between the environment and way of life of time past (to which our species is adapted) and the present situation. The archeological record suggests that the human way of life goes back perhaps 5 million years. Agriculture and the subsequent development of urban populations and the industrial revolution occurred well within the last 12,000 years. Over 99 percent of human history was spent as "man the gatherer-hunter," and what we think of as "human nature" evolved to cope with social situations and problems arising from life in small groups with face-to-face interactions. Small wonder that many human beings find it hard to feel brotherhood with the millions of strangers who now share their world.

We have many lines of evidence to draw from, then, in reconstructing the past. The more we can learn about the evolutionary history and adaptations of other primate forms, the more we will know about the processes which shaped our own species. Monkeys and apes present an array of natural experiments in the adaptation of life in social groups to the demands of a wide variety of environments. The evolutionary process itself is both conservative and opportunistic. Adaptive change involves the elaboration and modification of preexisting systems—it never starts over with a clean slate. In other words, a human being is a human being partly because he or she is descended from an ape and not from an early form of lion or dog. We can understand the whole of human nature if we appreciate both the ways in which we are like other forms of life and the ways in which we are unique.

A BRIEF HISTORY OF THE PRIMATES

The order Primates is very old, with its first representatives appearing among the early mammals around 70 million years ago. These first primates were quite different from the ones found today and can only be recognized by details of their teeth and anatomy of the skull. In this early period, which was before the evolution of true rodents, a number of primate and insectivore groups were living the way of life (ecological niche) which rodents later occupied. They were small creatures with small brains and long snouts, and they climbed using claws instead of nails. They probably skittered about in bushes and low trees or on the forest floor, and may have been mainly active at night.

The following period, about 50 to 60 million years ago, witnessed a great proliferation and geographical expansion of the primates. In fact, this was their heyday—after this time primates were relatively insignificant as an order until less than 12,000 years ago when one of its species, *Homo sapiens*, domesticated animals and plants and peopled the earth. The great proliferation and expansion of primitive primates was due to an adaptive radiation based on a new way of moving about, finding food, and relating to the environment. This new way was a locomotor adaptation—climbing by grasping instead of by using claws. The adaptation involved some important changes in the anatomy of the hands and feet: fingers and toes lengthened, their tips became fleshy, ridged pads for grasping, and the primitive claw thinned out and flattened into a nail which served as backing for the grasping pads. Today all the living primates have nails and use the grasping hand when they climb (although a few have retained one or two claws for other purposes on relatively unimportant fingers and toes).

There is argument today over what was so successful about the grasping hand and foot for the early primates (Jolly 1972). The best suggestion seems to be that these early primates began as hunters of insects and small vertebrates (Cartmill 1974). Instead of hunting at night and using the senses of smell, touch, and hearing, they hunted by day and used vision to locate their prey. The use of vision to locate fast-moving prey demanded other behavioral changes. Most mammals use their snouts and jaws to acquire and break up their food before eating. But visual hunters cannot afford to do this because they would have to lose sight of their victims just at the crucial moment of capture. Instead, the primitive primates began to catch their prey with their hands and to use their hands to break up their food into bite-size morsels. And so begins the story of the evolution of the human species. Selective pressure began favoring individuals with grasping hands and good hand-eye coordination. This was the first step along the way to the evolution of a species dependent on the use of tools for its living.

These early primates are called prosimians, and there are a few surviving species living today in protected niches where they do not have to compete directly with their more advanced relatives, the monkeys and apes. These survivors are found living in Africa and Southeast Asia where they are nocturnal (active only at night), filling niches very different from monkeys and apes, who are diurnal (active in the day). Other prosimians live on the island of Madagascar, where there are no other primates aside from humans. Although these prosimians are very diverse, they do hold some characteristics in common. They share the basic primate adaptation of nailed, grasping hands and feet. Their eyes tend to have at least partly overlapping fields of vision. This means that the eyes are rotated from being placed at the sides of the head to a more forward position. Overlapping fields of vision aid in depth perception, a great advantage to visual hunters but bought at the price of not being able to see in all directions at once, an advantage in spotting potential predators. Prosimians have relatively long snouts and wet noses which attest to the fact that the sense of smell is still very important to them. They also have long jaws compared to monkeys because they use them in a cropping rather than a grinding motion. During their heyday the prosimians must have lived over much of the world; their fossils come from Europe and North America, and their survivors are found in Africa and Southeast Asia.

Some of these primitive primates had other descendants but these descendants are called monkeys and not prosimians. Monkeys represent a second major radiation in the history of the primate order which involved changes in the special senses, in the use of the jaws, and in social life. This radiation is less well documented in the fossil record partly because the behavioral changes do not fossilize. However, we can use the comparative method to help in reconstructing what must have happened, although we cannot always say exactly when and where. We do know that by about 30 million years ago all of the major groups of Old World monkeys were represented in fossil beds in Africa. The process of evolving a monkey way of life must have begun at least 20 million years before that. Monkeys have fully overlapping fields of vision and unlike most mammals, they can appreciate color. Their senses of hearing and smell are comparatively reduced. When researchers are watching the behavior of monkeys and apes, they can feel confident that the sensory world of their subjects is essentially within their own reach, whereas in studies on dogs or rats the researcher knows that his subjects are responding to sounds and odors in the environment that he cannot appreciate without special instruments. Touch is also a very important sense for monkeys and apes just as in humans, but instead of using a wet, mobile nose and tactile hairs around the mouth, the primates use their hands.

Monkeys use their hands even more than prosimians, which is one of the reasons why their vision is fully stereoscopic. The original primate grasping hand with opposable thumb and nails became further modified as it began to take over many of the functions of the face and jaws. The monkey is a hand feeder, not just a grasper, and so each finger must move independently as he performs tasks such as the plucking of fruit, removal of a hard shell from a piece of fruit, breaking it up into pieces, and finally bringing it to the mouth for chewing. The face of a monkey became less important because it was no longer used as a tactile organ; the sense of smell was reduced and the nasal passages took up less space, and the jaws were used far less for capturing and breaking up food. Monkeys then have short, deep jaws suitable for grinding and chewing movements, but no longer useful for plucking or cropping like the long, narrow jaws of prosimians and many other mammals.

Today there are two major groups of monkeys, the New World and Old World monkeys (Fig. 1–1). As you can see in the diagrams, there is disagreement as to when the two groups separated from each other. In both cases, however, there is agreement that the New World monkeys are less closely related to the apes and humans than are the Old World monkeys. However, the New World monkeys are interesting because they present a wide, adaptive array of species which filled up many of the arboreal niches in the New World. They offer many interesting parallels in anatomy and behavior with the Old World monkeys and apes which help in understanding the significance of particular adaptations.

One of the most interesting aspects of behavior of monkeys in general is their commitment to life in social groups. As you will see in the following chapter, they have evolved many different ways to organize themselves socially, but nearly all of them have adopted life in social groups as the major adaptive stance.

About 25 million years ago a third major adaptive radiation began among the primates. It occurred in a group now known as the apes. The first apes, from about 10 million years earlier, can be recognized from fossil beds in Egypt. However, they are only recognizable by details in their teeth and it is probable that they did not actually evolve a distinctly ape way of life until as late as 20 million years ago. Before that they were essentially monkeys in their life ways even though their teeth were a little different.

The modern apes are set apart from monkeys because of far-reaching differences in their locomotor patterns which remodelled the entire body. As was mentioned earlier, monkeys are four-footed animals (quadrupeds). Even in trees they walk on

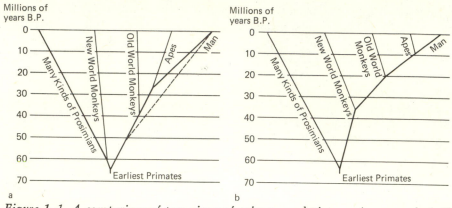

Figure 1–1. A comparison of two views of primate evolution. a. A commonly held view of primate evolution. According to this view, the New World monkeys are derived from North American prosimians, and their similarities to Old World monkeys are due to parallel evolution. Man has a separate ancestry for some 25 million years, or, according to some individuals, for much longer than that (shown by dotted line on the figure); b. This view of the primates differs from that shown opposite. New World and Old World monkeys have a long period of common ancestry after separating from primitive prosimians. The New World monkeys reached the New World, perhaps on rafts, crossing the sea from Africa some 35 to 40 million years ago, before the South American and African continents had drifted far apart. Man is shown separating from the African apes less than 10 million years ago, after sharing a long period of common ancestry with the apes.

top of branches similar to the way they walk on the ground. Apes are "brachiators"; that is, rather than stand on top of a branch and feed, they prefer to hang from it by one arm and feed with the other (Fig. 1–2). Brachiation has definite advantages in that it permits a relatively large animal to move out on the end of branches and hang, sometimes distributing his weight among various branches nearby, without having to worry about maintaining his balance over one branch. The changes in the anatomy of the arms and trunk which permit this behavior are far-reaching. The shoulder joint of the ape permits a 180 degree rotation as in humans instead of mainly in a single plane, as in quadrupeds. There are special modifications in the elbow and wrist joints to permit hanging by one arm and slow rotation in a circle while feeding. The flexor muscles of the arm and hand are so powerful and important that, when an ape tries to walk on the ground quadrupedally, he can no longer flatten out the palm of his hand and must walk on his flexed knuckles instead (see Fig. 1–2). The ape is no longer a true quadruped and does not run and jump from tree to tree. As a result of this the lower back of apes is very short and they do not have tails to use in balance for jumping.

The apes were very successful for a time. In fact, around 20 million years ago there were many more apes than monkeys both in terms of numbers and variety, but by 10 million years ago their numbers were reduced and monkeys were far more plentiful. Today there are only four species of ape left and three of these are

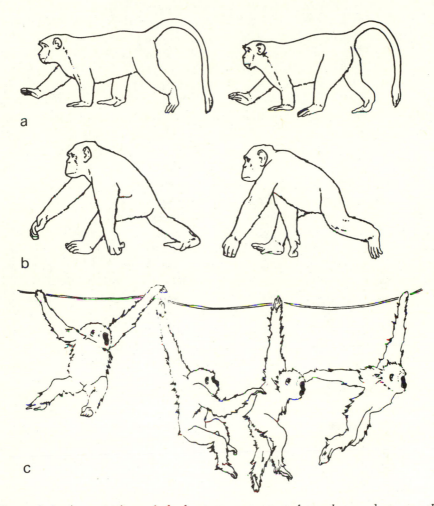

Figure 1–2. A comparison of the locomotor patterns of monkeys and apes. a. the quadrupedal walk of an Old World monkey; b. the secondary quadrupedalism of a knuckle-walking ape; c. brachiation or arm swinging of an ape.

threatened with extinction. There are the African apes, the chimpanzee and the gorilla, and the Asiatic apes, the orangutan and the gibbon. One might say that as a group apes were not very successful: after a brief flurry their numbers and kinds dwindled away and their geographic distribution shrank to very small areas. However, one species of ape was very successful and its descendant, the human species, eventually came to be the dominant species in most parts of the world.

The most recent radiation in the primate order, the one which led to *Homo sapiens*, was hardly a radiation by usual standards. It began with only two or three separate species perhaps 5 million years ago. These species did not increase in number and variety as do most radiations, but instead, after several million years, focused down into a single genus and species which moved into a wide variety of geographical and ecological zones without further speciation. The earliest known forms of true humans, the Australopithecines, are truly the once "missing link" between apes and humans (Le Gros Clark 1967). Today the Australopithecines are a very well-known group; representatives have been found in many different parts of Africa and there are some indications that they may have lived in the Near East and Southeast Asia as well. Archeologists have good collections of skulls and parts of skeletons of both sexes and adults and young. They have collected tools of the Australopithecines, studied their techniques of manufacture, and noted signs of wear on the working edges to learn how they were used. Australopithecine camp-sites have been found and careful analysis of waste flakes from tool manufacture and remnants of meals have been made. Even though the Australopithecines are the earliest known men, their way of life and the way they looked are better docu-mented than some of the subsequent stages in human evolution.

The Australopithecines look very much like a "missing link" should look al-though, when they were first discovered, many scientists found it hard to accept their humanity. Their most outstanding features are the nearly human pelvis and legs, the small, ape-size brains, and a clear commitment to tool making and using. The first discoveries of Australopithecines did not include undisputed tools and many researchers found it impossible to believe that a bipedal creature with a brain the size of an ape could possibly be making tools. Today this is no longer questioned because many stone tools have been found with Australopithecine bones in East Africa and because we now know that wild chimpanzees make and use tools, sometimes even crude bashers of stone.

Even after the Australopithecines were finally accepted as the first humans, scientists continued to think of them as representing a relatively brief, transitional phase in the evolution of man. Their reasoning was that tool using was such an adaptive behavior pattern that, once a species really got into it, there must have been tremendous selective advantage for intelligence and ease of learning, so that the human brain must have expanded very rapidly. However, it seems now that this was clearly not the case. As more and more finds are made, they are pushing back the first Australopithecines earlier and earlier in time. Australopithecine sites in Ethiopia range from 2 to 4 million years ago and other fragments come from Kenya which may go back as far as 4.5 and 5 million years ago (Pilbeam 1972). The most recent sites date around 1 million years ago, which means the Australo-pithecine stage of human evolution was no brief interlude but a long-term, stable pattern of life. If we take 5 million years (a conservative estimate) as the earliest possible date, then we can say that a minimum of 80 percent of human history was spent in the Australopithecine stage of evolution.

We will come back to the Australopithecines again during the course of this book, and the final chapter is devoted to discussing their adaptations and how and why they evolved. By the time the reader gets to that chapter we hope he will have a good understanding of the ways in which the primate background set the stage for the evolution of a distinctly human way of life.

TOPICS FOR DISCUSSION OR PROJECTS

1. Think about the pressure of natural selection on mankind in the future. There are major changes occurring in the natural world as a result of modern technology. Do you think that new physical, social, or emotional qualities may prove more adaptive in the future?

2. What do you think the effect of birth control and family planning will be on the process of natural selection?

3. The geneticist Dobzhansky has suggested that in the ideal democratic society of the future genetics will play a larger rather than smaller role in determining the course of life for the individual. Discuss.

2

The Social Life
of Old World Monkeys
and Apes

INTRODUCTION

The most striking feature of our primate heritage is life in social groups. The higher primates, that is Old World monkeys, apes, and humans, all solve the major problems of existence in a social context. For many animals mating, feeding, sleeping, growth of the individual, and protection from predators are matters for the solitary individual to solve, but for the higher primate they are most often performed in a social context. In fact many members of our order may well have been committed to this way of life for as long as 30 million years. This long-standing and also highly diversified commitment to life in social groups set the stage for the later evolution of the human pattern, with its total commitment to group life and the rapid expansion of cultural and social traditions of behavior.

Since the higher primates have been living in social groups for millions of years and since they have also occupied a wide variety of ecological niches, it is not surprising that we should find many different patterns of social groupings. Nevertheless, it is possible to make some very significant generalizations about the social groups of Old World monkeys, apes, and humans which hold true for most if not all of their member species (Eisenberg and others 1972). Typically the social groups of the higher primates include individuals from both sexes and from any stage of development from birth to old age. Furthermore, group composition varies minimally during the annual cycle. This kind of social grouping is very different from those typical of many mammals where males and females often do not associate outside of the breeding season and adults do not live with their young after they have reached puberty. In fact, the only other major mammalian group to develop similar social systems is that of the group-hunting carnivores such as wild dogs, wolves, and lions. The higher primates also have a relatively long period of development and life expectancy. Puberty for monkeys comes at the age of two or three and for apes at eight or nine. It is not uncommon for the larger monkeys to live 20 or 30 years, and apes have been known to live into their fifties in captivity.

These few basic facts—life in a relatively stable social group composed of long-lived individuals of all ages and both sexes—have major and far-reaching effects. They mean that young monkeys and apes are born into the company of individuals who will be important to them for the rest of their lives. In fact, a young infant may well grow up, reproduce, and die in the company of the same animals. There are

12

always some individuals who change groups but even they are likely to join with some close neighbors and their total social field will only have shifted slightly. This long-term stability of basic social relations has enabled the higher primates to develop social systems based on the membership of individuals as individuals and not as just representatives of a particular age–sex class. In other words, if you see monkey A interacting with monkey B you are not simply seeing an adult male interacting with a juvenile male. Rather you are seeing two individuals with special personalities, individual social histories, special social relations and alliances, and a history of past interactions with each other. In such a situation the juvenile male will respond to such vital information as to whether the adult male has an easygoing or irritable personality, if he recently joined the group or has lived with it for many years, and whether or not he is popular with the other adults in the group. For his part the adult male will respond to such facts as whether or not the young juvenile belongs to an alert and dominant mother and if she or other close relatives are nearby at the moment. Ultimately this means that higher primates must be able to make constant assessments of the social situation before acting. In evolutionary terms life in these kinds of social groups puts selective pressure on the ability to evaluate social situations and to develop a reliable social memory. Long before humans ever appeared there had been millions of years of natural selection for social intelligence, for brains large enough to make these many complex discriminations between individuals and situations and to store such information learned over a lifetime spanning several decades. It is not surprising then that the brains of monkeys and apes are proportionally large compared to most other mammals or that they seem to display so much humanlike intelligence when we watch them in the zoo.

It is impossible to fully describe the wide variety of social systems observed during the past 15 years of primate field studies. Nevertheless, five basic axes of social organization emerge which seem to represent behavioral potentials of most higher primates. They are like themes which are woven together to form a pattern unique for each species. Sometimes certain of these axes will be emphasized in meeting the challenge of a particular environmental problem confronting the species or perhaps only the social group. Sometimes, too, a particular social group may have a social tradition in which one of these themes may dominate whereas its neighbors will have another. The five basic axes of social organization are dominance, the mother–infant bond and the matrifocal subunit, the sexual bond between males and females, the separation of roles between adults and young, and the separation of roles by sex. The remainder of this chapter will discuss each of these axes in turn in an attempt to describe the variety of social systems found among the higher primates and how these systems meet particular environmental challenges.

DOMINANCE AND DOMINANCE HIERARCHIES

Dominance was one of the first axes of social organization to be described for nonhuman primates, probably because it is relatively easy to observe in the interactions of ground-living forms such as the baboons and macaques. In fact, many of the first studies on primates made it sound as if the dominance hierarchy were synonymous with social organization for nonhuman primates. Today, after many species have been studied in much greater time depth, it is clear that dominance is only one of a number of principles of organization which lie within the behavioral repertoire of the higher primate (Bernstein 1970; Chance 1967). How significant a role dominance plays and how structured dominance relations are varies from group to group and species to species.

Dominance hierarchies clearly have an important adaptive value. They allow individuals in a society to predict the outcome of an interaction when two animals compete for a scarce item in the environment, be it food, space, an estrous female, or only a safe, comfortable seat. Ultimately, hierarchies are based on the ability of one animal to physically dominate another in a fight, but this ability is based on a very complex set of factors such as the strength, age, health, motivational level, and social alliances of each animal. Through experiences which begin shortly after birth and continue through life, a young monkey learns which animals he can dominate and which he cannot. It is clearly not adaptive for individuals living together and having many daily contacts to have to fight every time an issue arises. It is far better for both the dominant and the subordinate to avoid the fight if the outcome of that fight is truly predictable. Besides being a waste of energy and socially disruptive, fighting can lead to infected wounds or tetanus, which could ultimately cause the death of an animal even if he was the winner of the fight. Instead of fighting every time, then, group-living animals have developed a substitute—threat behavior. Threats serve two important functions. First, they remind the subordinate animal of a previously established relationship, perhaps one founded on a single decisive fight years before or one slowly developed through years of interactions. Secondly, threat "displays" are mirrors of the motivational state of the signalling animal. They show exactly how aroused he is and how confident he is of winning should a fight develop. These threats can be very subtle and low in key such as a cold, unwavering stare, or they can be very noisy and showy such as the throwing of sticks or shaking of branches.

In the long run, then, dominance promotes tranquility even though it may be based on force. We so often mistakenly think of dominance as involving uncontrolled or unjust aggression, expecially when we think in terms of human relations, that its adaptive value is forgotten. Dominance is one effective way to organize social interactions, and unorganized social relations can be chaotic. Inconsistent assertions of dominance are especially upsetting. This is well demonstrated in small groups such as the human family. In situations where adults are very inconsistent in their assertions of dominance over children, serious behavior problems often develop. For example, healthy, happy children are produced by many different kinds of child-rearing techniques which range from very permissive to highly authoritarian depending on the traditions and values of the society. Trouble seems to most often develop when parents are so inconsistent in their dominance relations with their offspring that the children are unable to predict their parents' responses in a given situation. Inability to predict the outcome of such important interactions creates constant anxiety which can easily develop into behavioral and emotional problems.

The ability to predict the behavior of others is basic to the evolution of complex social systems. There is almost an emotional need for this in primate species. For example, in the laboratory when a new group is formed of either baboons or macaques, two relatively dominance-oriented kinds of monkeys, these strangers will interact at a very high rate until the major features of a dominance hierarchy emerge. Only then will they more or less relax and put more of their time and energy into other kinds of behavior like feeding, resting, and grooming. Ultimately, individuals who are unable to appreciate the major features of a dominance hierarchy will lose out in the selection process. This selection exists at all levels of the hierarchy and does not simply favor individuals at the top at the expense of those at the bottom. For example, a dominant male baboon may not allow a subordinate male to mate with a female in estrus. However, this does not mean that the subordinate will not mate with her. It simply means that, if he is to mate with the female, he must have the brain power to bide his time until a rock or bush are between them and the dominant male.

The most typical dominance hierarchy among primates is a simple linear one in which monkey A dominates monkey B who in turn dominates monkey C, and so on. This kind of hierarchy is typical of other group-living mammals and is very common among monkeys, especially those that live in relatively small groups of 10 to 20 individuals, in which there may be only a few members of each age–sex class. However, some monkey groups number up into the hundreds and among these another form of hierarchy may emerge in which a small number of individuals form an alliance against others. DeVore (Hall and DeVore 1965) has described this kind of alliance for the savannah baboon. In one of the troops that DeVore studied he found six adult males, all large and about the same size. These six males were of very different ages and physical condition, ranging from the young adult Mark, 3–5 years old, to Dano, a late, prime male about 12–18 years old (Fig. 2–1). There was even one very old male, Kovu, whose canines were broken off into stumps. Kovu

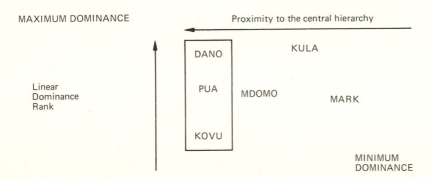

Figure 2–1. The central hierarchy. The adult male dominance hierarchy among a troop of savannah baboons. A male's position is a combination of his abilities as an individual (his "linear dominance rank") and his ability to enlist the support of other males in the central hierarchy.

could well have been aged 20 or more. These six males dominated the social life of the troop, which numbered 31 animals.

The adult males play a very significant role in the life of a savannah baboon troop because the safety of the troop may well depend upon them. Most primate species live either in trees or within sprinting distance of trees should danger threaten. Savannah baboons, however, sleep in trees at night but in the day move out into treeless grassland in search of food. The potential predators are many: lions, cheetahs, wild dogs, and even hyenas. In this ecological setting solitary baboons do not exist. All individuals including the adult males depend on the group itself to discourage predators.

Among most primates who live in or near trees, the response to the sighting of a hunting predator is the same, a pattern which, when seen among humans, is most aptly called "devil take the hindmost." In that case the adult males often beat the females and juveniles to the safety of the trees because they are bigger and faster. Among savannah baboons in open country the response is different. When an alarm call is given and the source of the alarm noted by all, the females and young will flee to safety but the mature males will hang back and may even form a line threatening the predator until the weaker members of the troop are safe. It is unlikely that these males would do actual combat with predators, but they probably would not be called upon to do that anyway. The highly visible show of force is enough to discourage any sensible predator. This role of defender calls for large, heavily muscled, aggressive males, well-armed with long, sharp canines (Fig. 2–2). These same males must also cooperate and get along with each other if they are to function as group protectors.

DeVore found a simple linear dominance hierarchy among these six males if he created situations in which only two males at a time had to compete for something highly prized, like peanuts. The hierarchy was as follows: Kula⟶Dano⟶ Pua⟶Mdomo⟶Mark⟶Kovu. However, DeVore noted that knowledge of this hierarchy did not help him very much in predicting the outcome of everyday interactions. He found, for example, that it was very difficult to get certain males, like Kovu, into a situation where they had to act alone against another male. In fact, he found that three of the males, Dano, Pua, and Kovu, were constantly together—feeding, traveling, or resting—and that they essentially belonged to their group not so much as individuals but as members of a social alliance or coalition. Together these three males formed what DeVore called the "central hierarchy" which was indomitable vis á vis the other males, who rarely formed alliances. This central hierarchy controlled the life of the troop because they acted in concert and the combination of three large males acted as a social magnet when it came to determining the direction of group movement or when it was time to rest or feed. DeVore also found that it was very hard to get these three males to compete with each other, they would avoid situations that might provoke aggression among themselves.

The effect of this coalition forming the central hierarchy was far-reaching. As you can see in Figure 2–1, Dano was not the highest-ranking male in the troop but he enjoyed the prerogatives of that status because he had two faithful allies in the two elders, Pua and Kovu, should Kula try to challenge his authority. For their

Figure 2–2. Canines: the aggressive weapons of an adult male baboon. Mark, a young adult male, displays his unworn teeth in a tension yawn. The upper canines are not fully erupted.

parts, Pua and Kovu gained in status through the coalition. Both males were old and had broken canines reduced to mere stumps. Alone neither could really enforce a dominant position but, as members of the central hierarchy, they enjoyed high status as long as they kept near Dano, which they nearly always did. Even the mating activities of the troop were influenced by the central hierarchy. Although Kovu ranked a very low sixth in the adult male dominance hierarchy, he ranked second along with Mdomo in the number of copulations he completed with estrous females during the mating season.

It is not always clear what history lies behind the rank that a particular individual may occupy. It may be that some males such as Kula never get to belong to a central hierarchy because their personalities do not permit them to cooperate with others in the group. Or perhaps Kula recently joined the group and, as a stranger, was relatively unpopular even though he was respected because of his high dominance rank. The role that experience during development plays in establishing dominance rank will be discussed in the following section on the mother–infant bond. For the moment it is enough to say that for many primates, the most important single factor in determining dominance rank is social alliances based on family ties. However, it is also clear that individuals, and especially some adult males, can achieve rank through special efforts of their own.

A good example of the achievement of rank can be found in the case of Mike, an adult male chimpanzee, studied by van Lawick-Goodall (1971). Chimpanzees rarely do injury to each other although their aggressive displays are noisy and involve a great deal of motor activity and excitement. Such encounters are unlikely to get to the point that a chimpanzee is actually bitten or harmed, especially one that belongs to the same community as the aggressors. One of the reasons for this is that the aggressive displays of chimpanzees are highly developed and very impressive. In fact, the actual elements of the display are also prominant in the aggressive displays of their closest relatives, gorillas and humans. Most typically the chimpanzee display involves running through vegetation, dragging and throwing objects, crashing through trees, and often drumming. In gorillas the drumming comes in the form of chest beating but in chimps it is most often in the form of beating on trees and stamping feet. The throwing and dragging of vegetation is not so very different from a favorite display in human domestic quarrels, the breaking of dishes or other objects that make a satisfying noise when shattering. In the same way drumming and foot stamping are very common elements in human dances, especially those preliminary to battle, which are meant to show off the vitality of adult males. These behavior patterns are not exclusively adult male, however. Among juvenile chimpanzees some elements of the pattern are first expressed in play at about the age of three. Females, too, give these displays, and van Lawick-Goodall noted that old Flo, the highest ranking female, often gave the full charging display complete with dragging of branches and foot stamping. However, most often the fullest expression and the highest frequency of the display occurs in males or in highly dominant individuals.

When van Lawick-Goodall first began her study of chimpanzees in 1960 in the Gombe Stream Reserve, Tanzania, she was not able to observe a clearly defined dominance hierarchy. Much later, after having worked out her data on aggression, she came to the conclusion that a hierarchy did exist but that it was not always relevant in explaining social interactions. She also felt that her introduction of feeding stations for a five-year period toward the beginning of her still-continuing study may have induced the chimpanzees to work out their relative dominance ranks because of the incentives provided by the bananas at these feeding stations. Clearly the feeding stations increased the rate of aggressive interactions, a fact which eventually led her to give them up. Nevertheless, some kind of rough hierarchy did exist among the adult males, and in 1963 Mike, a young, rather small adult male, was nearly at the bottom. He was usually the last to gain access to the bananas and was never the center of a male grooming cluster. For a while he had been nearly bald because so many clumps of his hair had been pulled out during aggressive interactions. Suddenly however, during a period of four months, Mike's social status rapidly changed because of an innovation he had learned to make in the typical adult male display. At that time the feeding station was right near the camp where many objects and discards from camp life were lying around. From time to time many of the chimpanzees tried throwing or dragging these in their displays. Mike discovered that, if he could get together two or three empty four-gallon kerosene cans, he could bash them around and vastly increase the noise level of his display. Eventually Mike found that he could make a very impressive passage through the feeding area by running full tilt bashing and kicking at three kerosene cans which he kept bouncing along the ground ahead of him.

Van Lawick-Goodall witnessed one particularly telling episode which occurred early in Mike's climb up the hierarchy. One day a group of five adult males, including the top-ranking three, were lying near the feeding station grooming each other. Mike was sitting alone about 30 yards away, staring at them. After 20 minutes Mike got up and walked over to the camp and picked up two empty cans. He returned to the spot where he had been sitting and watching the group of males peacefully grooming. Slowly he began to rock from side to side and, as the rocking increased in vigor, he began a series of pant-hoots which mounted to a crescendo. At the crescendo he rose and ran at the group of males, the cans crashing before him. The males rushed out of the way and Mike disappeared down the trail, his cans bouncing ahead. Some of the groomers reassembled, but only minutes later Mike returned and swept again through the clearing with his cans. Needless to say, the other males were impressed. Slowly they came to Mike, greeted him with gestures of appeasement and submission and then began to groom him. Only one male kept away, Goliath, who until then had been the unchallenged top-ranking male. Eventually Mike was able to dominate even Goliath through repeated use of his fantastic display. After some time van Lawick-Goodall decided she had to keep the cans hidden because Mike was becoming a menace, throwing cans at both humans and chimpanzees. Three years later she put out one can to see what he would do with it. Mike was about 200 feet away when he first noticed it, but within ten seconds he raced toward the can and began batting it around. Van Lawick-Goodall saw one young male dash 40 feet up a tree to safety. He stayed there a full eight minutes after Mike had left the area before cautiously climbing down again. Obviously, even if Mike never got a kerosene can again, the young male would not easily forget the impact of that display.

It is important to remember that dominance is only one principle of social organization and even for the most dominance-oriented species, it is only one aspect of social life. It may be useful to think of dominance as one axis of social organization that is particularly useful as a behavioral response to stress in either the social or physical environment. Generally speaking, the more arboreal a species of primate, the less the orientation toward dominance. This may be because ultimately there is less need for special large, aggressive individuals for group protection. Also, members of the same species but different habitats show remarkable differences in rates of aggressive behaviors and dominance orientation. For example, the same kind of baboons studied by DeVore in the Kenya savannah also live in the forests of Uganda where they were studied by Rowell (1967). In the forests the baboons did not show highly structured dominance relations among the males; in fact, males were very loosely attached to social groups and often wandered off on their own or with neighboring groups. Variations in rates of aggression and dominance orientations between different groups of the same species of monkeys living in trees or on the ground, in forests or on the savannah, in the country or in an urban setting, all suggest that, when the environment threatens the group physically and some kind of defense of the group is needed, dominance will become an important factor in social organization (Paterson 1973; Rowell 1967; Singh 1969). Perhaps it relates in some way to determining which males will take social responsibility and play roles which require the dangers and advantages of leadership and group defense.

There is another situation in which dominance also plays an important role, when environmental incentives are scarce and concentrated. As was mentioned earlier, van Lawick-Goodall felt that she had accentuated dominance relations among the chimpanzees by providing them with bananas, something highly prized which they had to compete for daily. By doing this, she made it important for the chimpanzees to know who comes first. When she stopped the artificial feeding, aggression and dominance interactions tapered off until they reached a much lower level. Similarly, with primates living in cages where space is limited and avoidance difficult dominance often determines which animal will get the best seat in the cage or who will eat first. It is interesting to note that there are other ways to compete for food besides using dominance and spatial displacement. A hyena does it by trying to eat faster than the others in the pack (Kruuk 1972). Primates, however, most often settle issues over priority of access to incentives by dominance, even though there will be vast differences in the ways in which dominance is expressed and how structured the dominance relations will be.

THE MOTHER–INFANT BOND AND THE MATRIFOCAL SUBUNIT

Many of the early field studies of monkeys and apes suffered from an overemphasis on the behavior of adult males. Adult male primates are often large and their behavior conspicuous. Even though they constitute only a small percentage of the total membership of a primate society, usually somewhere between 10 and 20 percent, they often get more than their share of an observer's attention unless careful sampling techniques are used. From this bias in observations, it has been only a very short step to see the social relations of that small minority of adult males as equivalent to the social organization of the group.

Another source of this bias in the early field studies of primates came from the length of the studies themselves. Early workers felt that a full study should include the entire year so that seasonal variations in social behavior would not be missed. No one thought that a full year might not be enough to reveal the major outline of social organization in a monkey or ape group. Schaller's classic study (1963) on the mountain gorilla was considered a monumental scientific achievement and yet it involved only a nine month investigation of a single population and 458 hours of actual observation under very difficult conditions. Today some field workers even discount statistical data from the first hundred or more hours of observation because of the time it takes to get familiar with all the members of the society and the major patterns of interaction.

By the middle of the 1960s, after the first round of primate studies had been published, it seemed to some that the major axis of social organization in primate groups was dominance. Over and over again studies on very different species of monkeys and apes reported on the dominance relations of the group; most often some sort of linear hierarchy was reported with adult males usually ranking at the top, and females and young below them. Then a new flow of material began on data from three separate ongoing studies: van Lawick-Goodall's study of chimpanzees begun in Tanzania in 1960, numerous studies begun in the 1950s by members of the Japanese Monkey Center on the Japanese macaque, and long-term behavioral observation begun in 1956 on a rhesus macaque colony on Cayo Santiago, an island off Puerto Rico. These three study areas yielded a wealth of material on another major axis of social organization in monkey and ape societies: the mother–infant bond which ramifies through time into a "mother-focused" or matrifocal subunit. There is one thing that all these studies have in common—they are all long-term endeavors, some of which begin to approximate the actual life expectancy of an individual.

In many ways the most basic theme running through primate social relations is not so much dominance as it is the attachment between a mother and her offspring. This bond between mother and infant is primary in the sense that it is the first bond to be formed in the life of the individual, and also in the sense that the individual peculiarities of that bond set the tone for other bonds formed later in life (see Chapter 3 on learning). Social bonding represents the primary adaptive strategy of higher primates, so that it is not surprising that powerful emotional supports have evolved which lead the growing infant to form these bonds in the first place and to maintain them and form new ones during its life. As Hamburg (1962, 1963) pointed out some years ago, primate and human societies are not composed of neutral or rational actors. Rather they are formed of very emotional individuals, many of whom need each other on a day-to-day basis in order to survive. Hamburg noted that the higher primates react to a threat to major social bonds in the same way as they would react to a threat on life itself. Emergency-type, physiological responses, the same ones which prepare mammals to fight or to flee to preserve their lives, are aroused when higher primates perceive threat to a major bond. In humans these responses, which are felt as blood pounding in the temples, flushing, sweating, a tightening or sinking feeling in the stomach, and agitation, often occur in domestic quarrels or when learning of death or sickness threatening someone who is deeply loved. These responses are the result of millions of years of natural selection for individuals who have feelings about their social bonds and belong to their groups as emotional, not neutral, actors. Physiological processes make the individual feel that the breaking of major bonds is intolerable and must be remedied.

One of the reasons why emotional attachments have been selected for so consistently over so many millions of years is because primates have a very long period of development compared to many other mammals. For example, a female vervet monkey reaches puberty and assumes adult social roles during her third year of life and she probably will not have her first infant until she is three or three and one-half years old. She will have been completely independent of her mother in terms of feeding and locomotion since the end of her first year, but social adulthood and the responsibilities of raising her own infant do not come for two more years. This long period of growth and learning would not be surprising if one were considering a horse or other large mammal but a vervet female is a little monkey who might never weigh more than nine or ten pounds.

The first attachment a young infant will form will be to its mother. This attachment must occur at birth because, unlike so many other mammals, monkeys and apes do not leave their young in nests, dens, or holes in trees. Instead the newborn is kept in constant contact with its mother. In monkeys the infant is able to support its own weight and cling to its mother's fur from birth. Mother monkeys feed in trees, fight, flee from predators, and jump from tree to tree with the infant holding on. Infants who cannot or will not do this, (who do not have the motor or emotional capacity to cling), will be eliminated from the gene pool through falls or predation if they are left alone. In the great apes and humans the situation is less extreme because the infant is born even more helpless than a monkey and the mother must help in carrying it. Nevertheless, the emotional content of the bond between mother and infant is very similar in monkeys, apes, and humans; it is deep and enduring. Rowell's work (1972) on baboons gives a good example of this. In the course of some experiments she was doing with caged groups of baboons, she removed mothers from their infants when they were about six months old. The infants remained in their groups and were cared for by other females while their mothers were caged separately out of sight and hearing. Over half a year later the mothers were returned to their groups. As each mother was carried into sight of the cage, her infant, who had not seen her for over half of its life, began to give "lost infant" calls and, when each mother was put into the cage, her infant rushed to her arms and their former relationship was resumed.

Sade (1965) demonstrated that the closeness between mother and infant does not end with weaning or even with the assumption of adulthood. He observed a group of rhesus monkeys living in an island colony off Puerto Rico. Among rhesus monkeys grooming is a major form of social interaction. During periods of rest and relaxation a group of rhesus will break up into clusters of groomers who are either tending their own bodies or their neighbors'. Grooming is clearly a very pleasurable activity among many primates. It is a self-rewarding behavior pattern in that monkeys who groom seem to be just as involved and enjoying themselves just as much as the individuals who are being groomed. This activity is undoubtedly one of the major ways in which social bonds are expressed and maintained. Sade became interested in what social factors influenced the choice of grooming partners, and he found that the colony was ideal for this because genealogical records had been kept for a number of years preceding his study. He found that the overwhelming factor in choosing a grooming partner was a close emotional bond based on the mother–young tie. Even during the excitement of the mating season when grooming between mating pairs usually precedes and follows copulation, the frequency of grooming between close relatives was unaltered. For example, one fully mature male was observed to direct 40 percent of his grooming activities to his mother with no change when he consorted with estrous females. She, on the other hand, directed only 9 percent of her grooming toward him because he was only the eldest of her three sons. Sade noted that a mother usually received a larger share of her off-spring's grooming than the offspring received of hers—her orientation was to her entire family with special attention for the youngest. As the years pass and more infants are born, the mother has less and less time for each individual (some females had as many as six or seven offspring living with them), but the older siblings learn that they can easily find partners for grooming, resting, and feeding among each other, even if mother is too busy. In this way the original bond between mother and infant is never really broken because of the need for attention by each new sibling. Instead, the original one-to-one bond ramifies to include subsequent siblings and even generations among those animals which an individual monkey can count on for affection, comfort, and support. This ramification of the mother–infant bond to include many individuals belonging to two or more generations forms a special kind of subgroup common to many primate societies. This subgroup is called a matrifocal unit because ultimately the bonds that attract these individuals to each other are based on their commonly held focus of attention and emotion, a mother or even a grandmother (see Fig. 5–2). There is no such thing as a patrifocal, or father-focused, unit in monkey and ape societies because for the most part mating systems are promiscuous and the role of father does not exist. Special relations do exist between young animals and particular adult males, but these males are more likely to be an older brother or even a mother's brother, rather than a male with whom a mother frequently mates.

The matrifocal unit probably exists in all primate societies, but there is a wide range of variability in how important it is in the daily lives of individuals. There may be a tendency for matrifocal units to become very significant in groups where the survival rate of young animals is relatively high and where group size is big. When the matrifocal group is relatively large, at any one time there are always suitable partners for grooming, resting, or feeding and the individual does not have to look outside the family for a social partner. However, there are also species differences in the exclusiveness of the mother–infant bond and the resulting matrifocal subgroup.

Kaufman (1973) has reported on a fascinating series of laboratory observations on two closely related species of macaque: the bonnet macaque from South India and the pigtail macaque from Southeast Asia. Although very similar, these two species have major differences in social responses which greatly affect the social organization of the group. First of all, bonnet macaques like to be physically close whenever possible, so when resting or sleeping they cluster into a huddle. Pigtail macaques, which are just as social as the bonnets, prefer to maintain some distance between each other, (Fig. 2–3), the only exception being a mother with her infant, whom she jealously protects from contacts with other group members. Bonnet macaques are oriented toward the entire social group for body contact, comfort, and the expression of affection. Mothers, of course, play special roles to their infants, but other females and even males will show maternal and protective re-sponses. When Kaufman and his coworker Rosenblum (1969) removed a bonnet

Figure 2–3. (top) *A group of bonnet macaques showing the characteristic huddling posture.* (bottom) *A group of pigtail macaques maintaining space between individuals. The group is roughly similar in composition to the bonnet macaque group above, with one adult male and about three or four females with their offspring.*

mother from her group, her infant was cared for and protected by other group members. Sometimes the bonnet infant was virtually adopted by another female and, when his mother was returned to the group, he did not go to her (Fig. 2–4). In contrast, pigtail macaques have a very exclusive relationship between mother and infant and separation was very traumatic for both. The infant first showed agitated distress and then depression. When he tried to get comfort from other females, they generally ignored him and sometimes even harassed him. When the mother was returned, there was a dramatic reunion as mother and offspring rushed into each others' arms. Kaufman and Rosenblum found that, by the time the colonies were ten years old, there was a dramatic contrast between the two species in social organization. Bonnet macaques showed generalized or diffuse orientation toward the entire social group for grooming, resting, and sleeping partners whereas the pigtails

Figure 2–4. (top) *A bonnet macaque female with her own infant and two separated infants whom she adopted.* (bottom) *A depressed pigtail macaque infant showing characteristic posture and expression. He is completely disengaged from a mother and her infant nearby.*

showed a strong development of three-generational, matrifocal subunits including a mother, her offspring (both male and female), and the offspring of her daughters.

Unfortunately, pigtail macaques live in swampy areas with thick vegetation and long term studies on them in the wild have proven to be very difficult. Presumably, there is an adaptive advantage for pigtail macaques or the rhesus macaques (studied by Sade) to live in social groups which have two levels of orientation, one to the group as a whole and the other to the matrifocal subunit. In contrast, bonnet macaques and some other types of monkey like langurs live in groups where the mother–infant bond is deemphasized and individuals tend to orient toward the entire group. In langurs this diffuse orientation begins at birth when the newborn is passed around from female to female to be held and inspected. Undoubtedly both heredity and experience play important roles in the development of these differences. A pigtail infant, raised from birth by a relaxed, permissive bonnet mother, might have very different feelings about his mother and the social group than if he had been raised by his possessive, true mother. That he would act exactly like a bonnet macaque is unlikely, but it would also be highly unlikely that he would grow up to be a typical pigtail macaque either.

In situations where the survival rate of young animals is high and the focus on the mother–infant bond is strong, genealogies can become very large and include a number of generations. For example, the Arashiyama troop of Japanese macaques has been under observation since 1953 (Koyama 1970). In 1966 this group contained 16 genealogies with a total of 163 living members and another 23 known members which had either left the troop or died (Fig. 2–5). The top ranks of the group were held by three old sisters and their brother, and taken all together they and their living descendents totaled 39 individuals. In Figure 2–7 there are numerous cases of three generations living together as well as a few examples of four generations. Here membership in a genealogy determines most of the daily decisions that an individual is likely to make; for example, grooming, resting, sleeping, feeding, play, and traveling companions are most likely to come from within the genealogy, particularly if it is high ranking one. It is even much easier for a monkey of any age to learn new behavior patterns from a member of its own genealogy than from other, unrelated group members (see Chapter 3).

Both field and captive colony studies suggest that it is the matrifocal core which provides a primate group with stability and continuity through time. A field worker, returning after an absence of a few years, may only recognize his group through the adult females. Most of the males he knew may have dispersed among neighboring groups during his absence. Many different studies on both terrestrial and arboreal monkeys and apes indicate that it is the male of the species that is most likely to wander. Males begin to wander at puberty. They are not necessarily thrown out of the group, peripheralized, of low status, or subjugated subadults, but rather they are adventurers.

Females, too, change groups on occasion but at a very low rate compared to males. Koford (1966) found that over a four-year period at Cayo Santiago there were 151 changes between groups, 91 percent of which were by males. Reports also come from studies in the wild of female monkeys and apes changing groups, but always at a much lower rate than males. The Japanese data suggest that when an

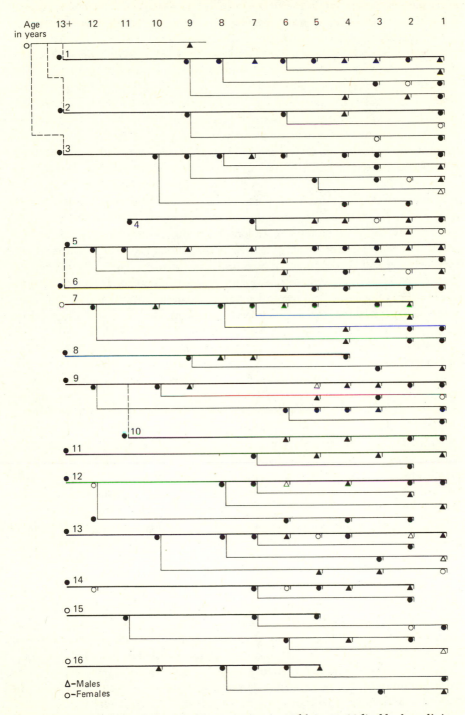

Figure 2–5. Genealogy of the Arashiyama troop in 1966 (n = 186); black = living, white = deceased.

adult female changes groups it may be a much more serious affair than when a male does. The female is likely to take her entire genealogy with her and she may in effect not change groups so much as she will establish a new one (Koyama 1970).

Among wild groups it is not clear how high the rate of male changes actually is, but in at least three terrestrial monkeys (vervets, rhesus macaques, and baboons) it is known to be very high and it is probably even higher in arboreal groups. For example, Gartlan (Gartlan & Brain 1968) found that only one male out of a total of six stayed with his main study group of vervet monkeys for the entire observation period of 13 months. Gartlan felt this was particularly significant because vervet monkeys defend their group's territory against inroads from neighboring groups. One would expect that if anything would reduce male wandering it would be territorial defense. On many occasions he saw males change between neighboring groups and within days vigorously defend their new territories against their former comrades.

There are some significant implications from this contrast between males and females in their rates of mobility between groups. Monkey and ape societies are attached to specific home ranges or territories even though individuals may not be. Terrestrial and semiterrestrial species tend to have large home ranges and to eat a wide variety of scattered and seasonal foods. This means that knowledge about the resources of the home range, and especially the sources of food and water during bad years, is likely to be held by the old females, the only members of the group that have spent their entire lives in the home range. In view of this, Rowell (1969) has suggested that the often-described traveling arrangements of open country primates may have been misinterpreted. Observers report that some adult males travel at the head of the column. These males play an "indicator" role as to what routes might be taken but they do not determine where the group will go. This decision comes from the center of the group where other males walk with adult females and young infants. Perhaps another way to see this is to say that the decisions are often made by the older females of the dominant genealogies which form the physical and social core of the group. The males with them are the ones most attached to them, and are often probably their brothers and sons. Some evidence to support this comes from a six-month observation of a small group of rhesus without an adult male (Neville 1968). The females of the group continued a normal daily round without disorganization. Eventually, after many months, an adult male from another group was allowed to join up with them but the daily round continued as before, uninfluenced by the presence of the new male.

Studies on the Indian rhesus macaque and on the Japanese macaque suggest that within the social group genealogies are often ranked. In fact, these studies have come up with the rather startling conclusion that, for most members of these societies, it is sufficient only to know the mother of the individual and its birth order among its siblings to know its rank in the dominance hierarchy of the group (Kawamura 1958; Sade 1967). Among these macaques a young monkey takes its rank directly from its mother, so that dominance relations between young peers are entirely determined by the relative ranks of their mothers. It is easy to understand why this should be so. Mothers respond to a threat to their offspring as if it were to themselves. If the threat comes from an animal ranking above the mother, she may snatch up her infant and flee or give placating gestures. If the menace comes from one ranking below, she will respond aggressively with threats and chases. A young monkey cannot help but learn both general social attitudes and the specific treatment due to individuals in the group from these experiences. At the same time other group members learn that a particular little monkey, no matter how physically insignificant, may have a powerful and aggressive mother to back him up. In groups where genealogies have grown large, a young monkey can count on not only his mother but also his maternal aunts and uncles, his older siblings, or his cousins when he needs help (Fig. 2–6). The Japanese have found that the high-ranking genealogies are likely to be the most cohesive, which is reasonable, since the higher the rank of the genealogy the more members are likely to directly benefit from their coalitions.

When these young rhesus and Japanese macaques reach full adulthood there is a clear split between the social experiences of females and males. Females continue in the routine of life which was established when they were very young. Sexual maturity and motherhood do not alter the bonds and rankings that have already been established. For males, however, the possibility arises at puberty of achieving a rank either above or below their genealogical position through dominance interactions and fights. Even so, it is clear that family connections have a strong influence on the status of males and that young males of the high-ranking genealogies are more likely to stay in their natal groups and maintain this high rank (Koford 1963). Because males may change their rank through aggressive interactions and because

Figure 2–6. A coalition between three brothers. Flo's three sons, Figan, Flint, and Faben, mutually support each other while facing Mike, the most dominant male. Faben, the eldest with raised hair, is responding to a charging display just given by Mike. His brothers ran over to stand beside him at this moment of high tension.

in a group they have a relatively high turnover rate, the male dominance hierarchy is very unstable compared to that of the females. For the males of a group, some changes in their hierarchy are likely to happen every year, but for females there is little or no change in their relative statuses even when groups split and genealogies break off (Koyama 1970; Missakian 1972).

Ranked genealogies have not been reported for all terrestrial primates, partly due to the fact that most field studies have been too short to determine the existence of genealogies. However, my own field observations of the vervet monkey suggest that this phenomenon is not confined to the macaque genus. It is also quite likely that in some species, like the chimpanzee and many arboreal monkeys, dominance itself plays a smaller role in daily life, so that the development of ranked genealogies would be unlikely even though the matrifocal subunit would still remain the major determinant of activities in the daily routine of sleeping, feeding, grooming, resting, and play.

In some monkeys such as vervets, langurs, patas, and geladas adult females readily band together against individual males in dominance encounters. It is quite likely that genealogical cores form the nucleus of such coalitions. For example, in the vervet monkey group I studied coalitions were easily formed against the top three males in the group if they offended the females by trying to monopolize some concentrated and prized food source or by frightening an infant. These coalitions were formed irrespective of relative dominance status so that even females from the lowest genealogies would chase a male who had made one of their infants scream. Males who had recently joined the group were more vulnerable to such attacks than were males belonging to high-ranking lineages but no males were immune. A typical sequence might begin with a female screaming and soliciting aid by giving rapid glances back and forth between the adult male and the females whose support she sought. The coalition would then attack, running and usually screaming at the male. The male would turn and flee, running as fast as he could until he reached the nearest tree or rock. He would then run up it and turn to face his chasers who threatened from below. After a minute or so of exchanging threats, the females would then move off, perhaps going back to the food or infant which was the cause of the problem. After that the male might be free to join them, but on their terms and not on his.

Such coalitions of females against a male never seemed to affect the rank of the male concerned. He was never caught or bitten and even a series of such encounters did not alter his individual rank. However, there is an effect on the male's behavior, if not his rank, because clearly his ability to bully subordinates is curbed. In particular, he learns to be very careful about frightening infants or juveniles, the single most effective way to arouse a coalition against him. Several times after an infant vervet had screamed, all the nearby adult males immediately left the vicinity even though they were innocent of frightening the infant. Apparently the potential of a female coalition against them created enough anxiety to make them leave. Although female vervets do not enhance their own individual social positions by coalitions, they do limit the potential social autocracy of the large, dominant males.

It is important to note that, although the matrifocal genealogy resembles a human family, there are two major differences. First of all, a matrifocal group is really a system of interactions based on affection. Monkeys and apes do not define relations so that individuals play roles which involve defined rights and obligations like brother, sister, mother, father, aunt, or uncle. Any behavior pattern which reduces the frequency of interaction between mother and offspring, such as the passing around of infants in langurs, works against the development of a strong genealogy. In other words certain behavior patterns can alter the sharpness of focus of a mother–infant bond so that the developing infant will seek social partners among members of his social group who do not belong to his genealogy. The second way in which a matrifocal group differs from the human family is that there is no father. The special binding of an adult male to the female or females with whom he mates is comparatively rare in primate societies and occurs only in very special circumstances.

The mother–infant bond which ramifies to produce a matrifocal subunit represents a major theme in the social organization of monkey and ape societies. This bond produces a behavioral potential which can be emphasized or deemphasized according to the social system and ecological setting of the group. It establishes a positive emotional base which endures and provides considerable stability to the society because it binds each succeeding generation to the group. Chance and Jolly (1970) point out that one of the most fruitful ways in which to analyze primate societies is from the perspective of what Chance has called "attention structure." In other words, who pays attention to whom is really a crucial social fact, a way to observe what must be going on inside the animals' minds during the course of daily life. In dominance-oriented societies subordinate animals are constantly aware of the presence and activities of those which are dominant. This awareness and anxiety on the part of subordinates is not equally reciprocated by dominant animals, who often seem oblivious of others. The matrifocal unit too represents a kind of attention structure but very different from one based on dominance. Attention is reciprocal among members of a matrifocal unit: mothers are concerned about their offspring's activities just as offspring are interested in what their mothers are doing. Matrifocality is a principle of primate social grouping which is different from but just as important as a dominance hierarchy, and in many primate societies these principles form crosscutting ties which bind individuals into the social group.

THE SEXUAL BOND BETWEEN MALES AND FEMALES

By now it should be clear that bonding between particular individuals or subgroups in primate societies occurs when there are special advantages to keeping specific individuals together. The same can be said about the sexual bond among the primates. When a special bond develops between sexual partners, we must look beyond the simple needs of impregnation of the female toward other needs formed by the social and ecological adaptations of the group. Generally speaking, among many monkeys and apes the potential of a sexual bond is relatively underdeveloped. Mating itself is usually promiscuous and the animals involved are more or less unemotional or matter-of-fact about their activities.

Take, for example, a typical mating sequence among the vervet monkeys in

Zambia. Infant vervets were born only during the few months of the year coinciding with the beginning of the rainy season. This was very adaptive because it meant that most of the early development of the infant came during the time of the year when food and water were plentiful. Mother vervets nursed their infants for a full year so that weaning came at the beginning of the following rainy season, again a good time for the infant to first be on its own in feeding. Mating was also seasonal and came mostly during a four-month period about six months before the beginning of the birth season. Female monkeys and apes come into estrus just like other mammals. This means that, at about the time during the menstrual cycle when the egg is ready to be shed, the female sends behavioral and scent signals that make her very attractive to males.

Male vervets tend to "check out" the females in the mornings during the mating season. A male will casually walk over to a female while she is feeding and sniff her vulva. He may then try to mount her. Sometimes the female is not quite ready to mate and she will turn and threaten the male until he leaves her alone. At other times she will accept his mounting by standing and bracing herself in such a way as to make it easy for him to mount. On other occasions when the female is at the height of estrus, she may approach a male and stand before him presenting her rear in an obvious signal for him to mount. Copulation in vervets is brief and unemotional. Sometimes a female will even continue to feed while the male is thrusting. The males rarely show any behaviors that suggest sexual jealousy or possessiveness. During the few days each month when the female is in estrus she will probably mate with a number of males in the group, even some who are not adult. Obviously later, when her infant is born, there is no possible way for any individual male to feel special "fatherly" feelings toward the infant.

The relatively simple, unemotional mating of the vervets has also been described for North Indian langur monkeys, forest baboons, and gorillas, and it represents a kind of base line for ground-living primates.

From this base other mating systems have evolved according to the needs of the species or group. For example, in the very dominance-oriented societies of some baboons and macaques access to estrous females becomes one of the prerogatives of dominance. A very dominant male may threaten other males away from an estrous female and try to keep her to himself in what has been called a "consort" relationship. During a consortship, which may last for several days, a male and female may stay together—resting, grooming, mating, and feeding a little apart from the others. Sometimes subordinate and juvenile males have opportunities to mate with an estrous female but not within view of the dominant male. They will have to wait until he is napping or until a rock or bush screens them. Generally, a female is not cooperative about maintaining an exclusive bond with a single male. She will mate with most males in the group and it is only the vigilance of the dominant male which may keep her apart from the others. In other words, the tendency for a bond to develop between a mating pair does exist in these monkey societies but the bond is a one-way affair; the male bonds to the female while she is in estrus and focuses his sexual activities on her while she, on the other hand, shows no particular fidelity to her partner.

There are two very good examples of primate societies in which the sexual bond has developed to form an exclusive grouping which in some ways resembles the human family. One of these is the hamadryas baboon, a desert-living species studied by Kummer (1968) in Ethiopia, and the other is the gibbon, a small tree-living ape studied by Ellefson (1968) in Malaysia. Both of these examples are informative because they show the way in which a sexual bond can develop to support a family-like grouping which serves important adaptive functions aside from mating.

Long before Kummer's study the sexual behavior of the hamadryas baboon was popularly known from travelers' tales and zoos. It captured the imagination of many social theorists as representing a primitive form of the family. Hamadryas baboons live in large troops which include individuals of all ages and both sexes. Within the troops there are special, very cohesive subgroupings which consist of a single adult male, one or more adult females, and their young offspring. These subgroups are called "harems" and their cohesiveness depends on jealous guarding and herding by the adult male. The male is ever vigilant regardless of whether his females are in estrus or not. He does not permit them to stray more than 10–15 feet away from him and, if a female should wander off or drop behind, he will punish her by grabbing her and biting the back of her neck. If another adult male without a harem should wander near, the harem leader will be even more alert and keep the females close to his side. Young adult males often begin their harems by "adopting" an infant female who is ready to be weaned. He will capture her and through neckbites force her to follow him away from her mother. From then on she will be by his side, sleeping in his arms at night as she did with her mother. A year or so later she will mature and begin mating with the male. During the course of his life a harem leader may add females to his subgroup by adopting new juveniles and by acquiring adult females from harems whose leaders have died. Leaders respect each other's exclusive rights to their harems but males without harems are ever alert to capturing a female from a harem or perhaps just mating with one who is in estrus. Just as with the consortships of baboons and macaques mentioned earlier, the female hamadryas has no sexual loyalty to her harem leader and will mate with any male she can whereas the harem leader strictly confines his sexual activity to the females of his subgroup.

Kummer was curious as to what the adaptive advantage might be for the development of this special subgrouping (the harem) within the larger group, the hamadryas troop. He noted that hamadryas live in semidesert areas in which food is scarce and widely distributed. This means that the members of a troop must scatter during the day if all are going to get enough to eat. When the hamadryas spread out during the day to feed, they move in small subgroups through open country, away from the safety of the cliffs where they sleep at night. This could put females with infants and juveniles in a very dangerous position in a country where dogs, lions, leopards, and cheetahs live. However, the harem system automatically assures that every female with her young has a large male protector at all times, regardless of whether the group is all together at the sleeping cliffs or dispersed to feed. The harem is not so much a unit adapted for mating as it is for foraging in an environment of scarce and scattered food resources.

The gibbons (Ellefson 1968) also represent a special case among monkeys and apes in which the sexual bond between a male and female is especially strong and enduring. Gibbons live in small groups which superfically resemble the human family because they are composed of a mated pair and their immature offspring. Each group controls a small patch of tropical forest which it jealously protects from neighboring gibbons. These territories are not large, but they contain enough food to maintain a small group throughout the entire annual cycle. When a young male or female reaches sexual maturity, it is gradually driven from the natal group through a series of aggressive encounters with the parent of the same sex, so that the group rarely grows beyond five or six individuals. In many ways this little group of gibbons is much like a pair of nesting birds which keep away competitors of the same species from the area around their nest while they are raising their young. This assures them of a food supply close to home which will be adequate to support their developing offspring. The gibbon "family" is a common pattern for small birds and mammals in the tropics and does not represent a predecessor to the human family.

In the past much has been made of the mating systems of the nonhuman primates compared to humans'. The old cliché of the dominant male who has either an exclusive harem or else first choice of an estrous female is often brought up and suggested as part of the human species' primate heritage. The field data, however, does not fit this description. In fact, there is no one primate pattern but a wide range of variation in form, from extreme promiscuity to highly exclusive harem systems, and casual mating to highly excited, emotional bonding between mating individuals. Perhaps the best generalization that can be made about primate mating systems is that they tend to be compatible with or even be a mainstay for the social system, which is itself adapted to meet the demands of the environment.

Some of the variation between groups is better understood when the evolutionary role of the female as well as the male is analyzed. For the most part male strategy is to get as many females pregnant as possible, without paying too high a price in terms of competition with other males. The strategy for a female is very different. She has no problem in getting pregnant; her problem is to raise as many infants as possible to maturity. For a female, males are a resource in her environment which she may use to further the survival of herself and her offspring. If environmental conditions are such that the male role can be minimal, a one-male group is likely (Rowell 1974). Only one male is necessary for a group of females if his only role is to impregnate them. This is why among tree-living monkeys the most common (but not only) form of social group is composed of one male with several females and their young. Tree-living monkeys do not have to worry about defense from predators in the same way that open-country primates do. In forests predators do not hunt in groups but rather by stealth, and the best defense against them is to be alert and to stay out on small branches which will not support a large animal. The role of an adult male is minimal in these circumstances and one is enough for a small group of females. As summarized by Eisenberg and others (1972), the number of males in a given group will depend on the advantage of their presence to the reproducing females. In general the more ground-living and open-country the adaptation, the larger the male role will be and the less the likelihood of a one-male mating system (the only exception being groups like the hamadryas where there has been selective advantage for a minimal-sized foraging unit). The more the male role expands with respect to responsibility for vigilance and group protection, the more the leader male needs to take on helpers and share the sexual favors of the females to ensure the survival of his (and the other males') offspring. This analysis from the perspective of the female primate does not deny that there is competition between males for the role of leader male in a one-male group or dominant male in a multi-male group because of the evolutionary advantage that role may carry in respect to impregnation of females. This point of view simply emphasizes that male and female evolutionary strategies can differ even though mutually complementary. Neither strategy alone can be used to give a complete analysis of social behavior. Both are equally valid and we ought to consider each one in trying to understand a social system and how it works.

Mating systems then must be understood in terms of adaptation to the environment in which these systems occur. They relate to the roles played by males and females in the total social system. The linking of a specific male to a specific female to form a bonded subgroup within a society is rare for monkeys and apes but does occur in situations with unusual ecological demands, as in the case of the hamadryas harem, which is really both a foraging and a reproductive unit. We may turn later to these same kinds of reasons for an understanding of the bonding of the male into the human family.

THE SEPARATION OF ROLES BETWEEN ADULTS AND YOUNG

In many ways the single most important adaptation of the higher primates (Old World monkeys, apes, and humans) is a pronounced division of roles between adults and young. One might almost say that upon this all else depends. Clearly a characteristic of these primates is the very long time that it takes to grow up, especially when relative body size is considered (Fig. 2–7). Obviously it takes years for an adult elephant weighing literally tons to grow, but it is surprising to find that it takes a three pound Talapoin monkey (the smallest of the Old World monkeys) three years to reach maturity. Growth here has to be in nervous tissue, the brain, and not in simply bone and muscle. The years of development before a primate reaches adulthood are spent in the protected environment of the social group, where the young juvenile has plenty of time to play and practice the social and physical skills needed in adulthood. This is the time for development of the brain through play and learning without which the special intelligence characteristic of the higher primates would be impossible.

The next chapter is devoted entirely to the topic of learning in primates, so the purpose of this section is to focus on the social setting in which that learning and development normally takes place. As Hall (1968) has pointed out, the context of learning is as important as what is learned. This is so because, without the proper psychological or emotional context, learning will not take place no matter how valuable or adaptive the learning might be to the individual. The single most important mechanism for learning in the higher primates is play and lots of it, and a playful state of mind is clearly inhibited by fear or anxiety. The young developing primate does not spend hours every day hidden in a nest or cave while the mother feeds. Instead he spends his life in the safety of the social group where both a

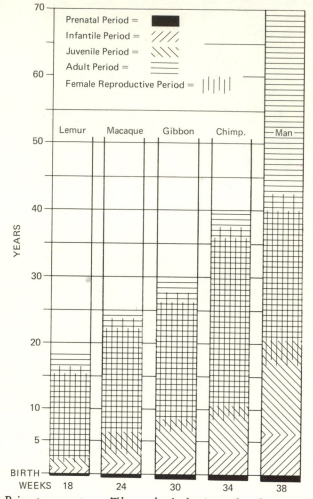

Figure 2–7. Primate age spans. The end of the juvenile phase is marked by the onset of reproductive maturity. Males of many primate species such as baboons do not reach full social maturity until several years after reproductive maturity is reached.

behavioral and psychological separation of roles between adults and young frees him to play for many hours every day. Because alert adults are always present, the young primate is free from anxiety and able to play with abandon in chasing and playfighting with his peers, to manipulate sticks or other objects with rapt concentration (in the case of chimpanzees and humans), or perhaps just to rest in the deep sleep so characteristic of the young primate after a stimulating day.

This physical and psychological safety of the young primate is provided by the social group because adults play protective roles towards young animals as a class when there is an external threat to the group. Field workers often report that, when a danger call is given and all animals rush to the trees, a young monkey can jump onto any group member and ride to safety. Perhaps this is why infant monkeys so commonly have very distinctive coat colors. For example, in many monkeys in which adults are gray such as langurs and vervets, infants are black, or in the spectacled leaf monkey, adults are black and infants are pumpkin orange.

As mentioned earlier, "devil take the hindmost" is the most common primate response to predators, with the exception that a mother makes special attempts to rescue her offspring. However, there are a few examples from both monkeys and apes where adult males take extra risk in threatening potential predators or intruders while females and young rush off. What usually happens is that the leader male or several dominant males will stand firm and face the danger, giving threats such as barking and branch shaking (baboons and macaques) or chest beating (gorillas). Most predators are impressed by this—they may not give up the hunt altogether but even a moment's hesitation can mean the difference between reaching the safety of trees or being caught and eaten. Most predators are quite capable of assessing the situation as to relative risk and they show marked preference for an easy kill. It is not worth it to receive a wound which may bleed heavily or become infected just for one meal—there is always other, less aggressive prey. Struhsaker and Gartlan (1970) report a remarkable sequence in which a young patas monkey was rescued by three adult males. A jackal had caught the juvenile at a waterhole and ran off with it screaming and struggling in its mouth. Three adult male patas who were nearby (only one of which belonged to the juvenile's group) gave chase and pursued the jackal until he eventually dropped the juvenile unharmed.

Dramatic incidents of defense and rescue of young by adults are comparatively rare. Far more important really is the day-to-day protection that the simple presence of alert adults gives to the young. In many species of monkey, especially ground-living ones, various adult males and females act as baby-sitters for a play group while the other adults feed. Perhaps a single adult male or a couple of adult females will be the focus for a play group of five to ten youngsters. The baby-sitter may be an individual who has finished feeding and is simply resting and gazing about or it may be an adult who for some reason tends to be both confident and vigilant much of the time. The presence of such individuals give youngsters a focus for their play and allows them the psychological freedom to throw themselves completely into their romping and chases.

It is important to remember that this separation of roles between adults and young in no way involves an economic division in which adults provide food for the young. Aside from the basic mammalian adaptation in which a mother nurses her infant until it is old enough to live on the adult diet, adults are not responsible for seeing that young have enough to eat. Adults feed themselves, and they know what is safe to eat and where to find various foods and water at different seasons of the year. Juveniles benefit from this knowledge through their association with adults. However, an injured or sick youngster still has to feed itself and get itself to water or it will die virtually before the eyes of other group members. Individuals who would risk their own lives in defense of the youngster are psychologically incapable of seeing its need for them to bring it food and water. Once weaned, then, young monkeys and apes must feed themselves even though dependent on adults for many years for protection. Part of growing up for primates, including humans, is a gradual loss of the completely carefree mind, one that plays with no thought to danger and sleeps deeply, often oblivious to sounds, even sudden noises. This change in awareness about the social and physical environment probably relates to changes in levels of arousal and an increased ability of the mind to pay attention to more than one thing at a time, as well as to actual changes in social roles.

In summary, the separation of roles between adults and young is a key behavioral adaptation in primate social groups. It creates the kind of situation in which young primates can spend many years of juvenile dependency while they learn and practice the social, intellectual, and physical skills they will need as adults. Without this proper setting for learning created by the social system, the larger brains and longer period of development of the higher primates would be wasted.

THE SEPARATION OF ROLES BY SEX

Whether or not there is a great difference in the roles played by males and females varies from species to species among the primates. In a number of primate species there is actually a minimal amount of role differentiation between the sexes beyond the obvious one in which females care for infants because they are equipped to nurse them and males are not. In the more ground-living species both males and females are very attracted to young infants, are very tolerant of them, and show protective responses toward them. In fact, it is not uncommon among the macaques and baboons for an adult male to "adopt" or take on as a protégé a young, weaned juvenile. When special relations between an adult male and a juvenile form, the adult will groom the youngster, cradle it in his arms when it sleeps, protect it in dominance interactions, and generally play a maternal or protective and solicitous role. In some cases in which the genealogies of the individuals are known, it turns out that the juvenile is often a younger sibling of the adult male, and he has increased his rate of interaction with it after it was weaned and their mother's attention turned to a new infant. A few cases of adoptions of orphaned, weaned infants are known for chimpanzees, baboons, and rhesus macaques. In these cases the infant was adopted by an elder sibling who was still closely attached to the mother before her death. There is no evidence to suggest that inexperienced females are any more fit than inexperienced males in caring for these infants and both sexes are motivated to try.

Among the more tree-living primates adult males are usually aloof from young animals—tolerant of them perhaps, but not drawn to them in the way that semiterrestrial and terrestrial males often are. In many of the highly arboreal species the infant is kept very close to its mother for a long period during its early development, and probably the need for any special relationship with adult males is minimal. As mentioned earlier, one of the common forms of social organization in tree-living primates is a group of one adult male, several females, and their young. The contribution of the male to the group's well-being is minimal and usually consists of impregnation of the females and occasional aggressive displays towards potential predators. In fact, the male himself is likely to be loosely attached or semiperipheral to the social core of the group. He more or less follows the females and young about, protecting them from other males and perhaps from predators. If the male should die or disappear, the females do not show any psychological dependence on him but rather continue their usual rounds unruffled.

In contrast, the roles played by adult males in ground-living and open-country primates are sometimes, but not always, correlated with great sexual dimorphism in body size, weight, muscularity, distributions of hair on the body, and size of canines. The adult male baboon is a prime example of this kind of dimorphism (see Fig. 2-2). His total body weight is more than double that of an adult female. His shoulders and neck are covered by a thick mane of long hair which helps protect these areas in fights at the same time that it adds to the impression of size and strength. His upper canines are long and razor sharp, constantly honed by grinding against specially formed teeth on the lower jaw. It is interesting to note that the sharp canine teeth of monkeys are very different from the sharp, cone-shaped canines of predators such as lions or wolves. The predator's canines are adapted for grasping and piercing whereas the primate's are developed for slashing, as with a knife. The primate canine is not conical in form but has a sharpened back edge so that the tooth can be brought down against the flesh of an opponent and then drawn toward the attacker in a combination stabbing and slashing motion. These teeth are uniquely evolved in the primates and represent adaptations for fighting and not for catching prey. In primate species with well-developed sexual dimorphism in body size, there is usually a parallel dimorphism in canine length.

However, there is no simple correlation between habitat and sexual dimorphism (Table 2-1). Several examples of extreme dimorphism in body size are correlated with arboreal life (proboscis monkey and the orangutan) whereas instances of minimal amounts of dimorphism are found in both tree-living and ground-living species (leaf-eating monkeys, chimpanzees, and humans). This is so because there is usually more than one way to solve a problem posed by the environment; the way chosen is likely to relate both to the evolutionary history of the species as well as the adaptive niche they are presently occupying. So, for example, humans and their closest relatives the chimpanzees have relatively little development of sexual dimorphism in regards to size and strength. In the course of evolution early humans must have moved out from the protection of trees into more open country with greater dangers from the group-hunting carnivores. However, they did not solve this problem as the baboons and patas monkeys did by selecting for adult males twice the size of adult females. Rather they responded in an apelike or chimpanzee fashion through the use of weapons and group defense.

In species where dimorphism is minor, there may be very little difference in the aggressive potential between males and females. For example, in the gibbon (the small ape that lives in mated pairs) the body and canine size of males and females is virtually the same. It is not surprising then to learn that female gibbons are highly aggressive and act as partners to their males when defending their territories against intruders. So, too, in the vervet monkeys of Zambia females joined in with the males in driving off neighboring groups who had invaded the territory. In these situations

TABLE 2–1   PERCENTAGE RELATIONS BETWEEN THE AVERAGE BODY
WEIGHTS OF FULLY ADULT FEMALES AND MALES IN OLD WORLD
MONKEYS, APES, AND HUMAN (AFTER SCHULTZ 1969:201).

Species	Body Weight Percentage (Female to Male)	Habitat
Indian macaque	69	semiterrestrial
East African baboon	43	terrestrial
Indian leaf-eating monkey	89	arboreal
Proboscis monkey	48	arboreal
Lar gibbon	93	arboreal
Orangutan	49	arboreal
Chimpanzee	89	terrestrial
Gorilla	48	terrestrial
Human	89	terrestrial

there seemed to be a differentiation between females carrying young infants and females who were not encumbered. The infantless females would join the males while the new mothers would hang back and watch. Sometimes in cases of mobbing predators like pythons or crocodiles, a female with a young infant would pull it off her chest and plop it down near a "baby-sitter" before running over and joining the others in threatening the predator.

It should be noted that in this discussion of role differentiation between male and female in primate societies there is no mention of major differences in economic or food-getting activities. With few exceptions the adult males and females in a group eat the same range of foods in roughly the same percentages, although there are instances in which very large males may have some advantages in feeding on fruits or nuts in thick shells because of their stronger jaw muscles. The only really important and interesting exception to this is in the case of active hunting for meat by baboons (Harding 1973) and chimpanzees (Teleki 1973). Many of the terrestrial and semiterrestrial and some of the arboreal monkeys show a marked preference for certain items of protein such as birds' eggs, grubs, fledgling birds, lizards, and large insects. If they should happen upon these delicacies in the course of their food quest, they will be eaten with relish. Beyond these, there are special instances among baboons and chimpanzees in which certain group members actually join together and cooperate in catching and eating a relatively large mammal such as a rabbit, a young antelope, or even a species of monkey. These cooperative hunters are usually adult males even though females clearly like to eat the meat and may even beg for a share. Although there is no indication that meat is a major part of the diet of either baboons or chimpanzees, (they probably could give it up altogether with no harm done to the diet), it nevertheless represents a very interesting behavioral potential. This is especially so when one considers that in humans the evolution of cooperative hunting by males is an integral part of the species' adaptive pattern. The final chapter of this book will discuss the evolution of the human pattern in greater detail. For the moment it is only important to know that in most primate species there is practically no separation of food-getting activities between the sexes.

CONCLUSION

Five major themes of social organization have been presented in the preceding sections: dominance and dominance hierarchies, the mother–infant bond and the matrifocal subunit, the sexual bond between males and females, the separation of roles between adults and young, and the separation of roles by sex. Each species and perhaps each group has its own unique combination of emphases on these elements which weave a collection of individuals into a social system. In primates it is the social system itself which enables members of the group to meet the demands of the environment (Kummer 1971).

Undoubtedly, as more varieties of primates are studied in more kinds of habitats other themes will emerge. For example, recent field work on the tiny talapoin monkey (Rowell 1973) suggests another principle of bonding—attraction between individuals of the same age–sex category. In talapoin society subgroups are formed of adult and near-adult males, adult and near-adult females, and young juveniles with an adult male baby-sitter. Infants, of course, stay with their mothers. These subgroups occur in both day and night living and are quite separate in terms of social interaction even though they travel together as a group. Outside the mating season adult males and females are rarely seen together and never seem to interact. This particular form of social bonding may turn out to be a curious exception in primate social organization found only in the talapoin, or it may be recognized eventually as another very common theme specially adapted to certain environmental circumstances. This chapter is not definitive but it does summarize the state of our knowledge today about the major axes of social organization in the Old World monkeys and apes.

By now it must be clear that humans are not also primates by chance. Many of the themes which bind together human society are shared by our closest relatives and existed long before we ever appeared on the scene. In many ways the social behavior of monkeys and apes in time past must have set the stage for the evolution of the human way of life. By understanding the social behavior and adaptations of our closest relatives we can far better understand how and why human beings are both like and different from other primates. Similarities and differences between species of monkeys, apes, and humans become meaningful when they are seen as part of an adaptive pattern uniquely formed in each.

TOPICS FOR DISCUSSION OR PROJECTS

1. Review some of the Case Studies in Cultural Anthropology (published by Holt, Rinehart and Winston, Inc.) with which you are familiar or read some of those listed at the end of this book. Can you analyze them using the five basic themes of social organization found in nonhuman primate societies? How do some of these societies differ in their relative emphasis on these themes? Are there other themes important in the organization of small human groups that were not found in nonhuman primates? Can you suggest some possible adaptive reasons for some variation between societies?

3

Social Traditions
and the Emergence
of Culture

BIOLOGY AND EXPERIENCE

Natural selection acts upon the entire organism: its anatomy, its physiology, the way it behaves, and its relationship to the environment and other animals. Learning is only one aspect of the behavior or potential behavior of a species. This is why the old scientific arguments over nature versus nurture, or heredity versus environment, are false issues. Learning ability, just like any other biological characteristic, has been shaped and directed through evolution to serve specific ends. Each animal species has evolved to learn the things that it has to know just as the emotions have evolved to make an animal want to do what it has to do. Probably learning is slightly different in every species and these differences represent adaptations to the particular circumstances and characteristic environment.

Perhaps the most basic of all primate learning involves emotional and social attitudes. In Chapter 2 and later in Chapter 5 we discuss how the original bonds between mother and infant set the tone for social bonds that an infant will form throughout life. The first social bond establishes the expectations of the individual for all social experiences in terms of how he feels about them and what he expects from them. Later he will form social attitudes learned through experience. He will find that he can dominate others in the same way as his mother because, in the beginning at least, she will be there to back him up in his social encounters. In the same way a youngster will learn to be afraid of certain things in his environment such as snakes, unknown males, or potential predators. The intimacy of the mother–infant carrying position, in which the infant is held in front of the mother touching chest-to-chest means that he can easily see her facial expressions and feel changes in breathing, heart rate, and muscular exertion. Laboratory experiments show that it is enough for a young monkey to see his mother give a startled reaction to a hidden snake in a box for him to nervously avoid the box in the future (Hall 1968). This chapter will not be concerned with the learning of emotional attitudes such as social bonding because they are covered elsewhere in the book. Instead we will focus on behavior patterns which are part of the traditional experience of the group and which help it to meet the demands of the natural environment.

What is learned, how and when, and under what circumstances are all part of the adaptive biology of a species. Various behavior patterns may be learned differently because of natural selection. For example, field workers have noted that one group of baboons may have a different variety of food items in their diet from that of neighboring baboon groups who live in similar home ranges (DeVore and Hall 1965). They have also noted that nowhere do baboons seem to eat all the foods available; one group may carefully avoid a particular fruit or plant that another group finds a delicacy. Hall observed baboons in South Africa who ate scorpions with relish after first rapidly rolling them on the ground to stun them so they could not sting while being eaten. Other baboons carefully avoided scorpions, showing no interest in them at all as potential food. These differences represent group traditions based on the fact that, among many monkeys, food items are learned through a very specific process. Obviously trial and error learning would be very inappropriate in the tropics where many plants, fruits, and small vertebrates are poisonous. Instead, a young monkey watches its mother very carefully, sampling bits of what she is eating. Studies show that young animals are very open to trying new foods, sometimes on their own, but usually items that they see other monkeys and especially their mothers eating. By the time Japanese monkeys are one year old they have already learned the basic diet of their group—perhaps up to 120 items (Itani 1958). Maturity brings increasing conservatism so that it is very hard to introduce new food items into the diets of adults of many primate species; sometimes it is so difficult that a captured animal will starve rather than eat unfamiliar foods.

There are also biological limits on the learning of a species, not so much in terms of intelligence but in much more specific ways. For example, van Lawick-Goodall (1970) has described the way chimpanzees "fish" in termite nests using long straws. Termites, a rich source of protein, are popular food items in Africa: they are eaten with great excitement by birds, monkeys, and other animals when they emerge from their nests for mating flights during the rainy season. The rest of the year they stay well protected within the concretelike shell of their nests. When the rainy season begins, the termites tunnel out to the surface. There are a series of mating flights and after each one the holes are sealed over again by the workers. Chimpanzees scratch off this thin covering at the end of the tunnel and insert straws. The termites rush to attack the intruding straws and grasp them with their jaws. The chimpanzees then gently remove the straws from the holes and nibble off the termites at leisure. By using the fishing technique, chimpanzees are able to eat termites throughout the entire rainy season whereas other animals can feast on the insects only during the infrequent mating flights.

Chimpanzees are not the only primates that like termites, and van Lawick-Goodall has seen baboons near the chimpanzees while they are at work, sometimes even catching and eating a stray insect. Although baboons are eager to eat them, they have never been observed trying to fish for termites themselves. The baboons do occasionally watch the chimpanzees termiting but not with the same intensity as do young chimpanzees watching adults. Baboons are certainly dextrous and intelligent enough (in a general sense) to learn termiting and yet they do not. Their failure to learn in spite of the opportunity to observe and motivation to do so, must relate to the fact that baboons are not accustomed to explore their environment using a probe of some kind. Chimpanzees in zoos and in the wild often use sticks or straws to poke things: holes, strange objects, or baby brother. The use of a probe to investigate the environment appears to be a part of the behavioral repertoire of the species which has the potential of being refined into sophisticated patterns like termiting through observation and practice.

THE CONTEXT OF LEARNING: THE SOCIAL GROUP

Among monkeys and apes learning is first and foremost a social act. The natural context of learning is the social group. Young monkeys and apes do not survive alone in spite of the fact that they can feed themselves after weaning. The lengthening of the period of juvenile dependency in monkeys, apes, and humans simply means that the time in which the skills and attitudes of adulthood can be learned has also lengthened, so that much more can be learned than in more rapidly developing forms of life. During this long maturational period the juvenile gradually learns the body of traditional knowledge held by the adults in the group. In particular he will learn emotional and social attitudes about the way group members relate to each other and the attributes, both dangerous and advantageous, of his environment. Most of the young primate's knowledge will be acquired through a process called "observational learning," in which the demonstrator animal is passive and makes no attempt to teach, and in which it is up to the observing, learning animal to be watching at the right time and to be influenced by the significant elements of the pattern (Hall 1963).

For workers who are familiar with human but not with monkey learning, it usually comes as a surprise to learn that female monkeys do not purposively "teach" their infants anything. Mother monkeys influence the behavior of their infants in three ways. First, they may physically restrain an infant if it is doing something which makes the mother uneasy, such as moving too far away from her or near an irritable adult male. Second, if the infant does something which is annoying to the mother, such as pulling her fur or nursing when it is not wanted, then the mother may actively punish the infant by slapping, pushing, or nipping it. Another way in which the female monkey influences the behavioral development of her infant is by setting an example for it: the infant watches the mother as she feeds, relaxes, interacts with other monkeys, or responds to danger, and it tends to make responses which parallel hers.

This inability of monkeys and apes to teach their infants actively is not due to their lack of language. For example, the teaching of manual skills even in man can be done without words, and in some cases language almost gets in the way because it is often difficult to talk about motor acts. Good, repeated demonstrations combined with grunts to emphasize crucial points are enough. Mother chimpanzees could teach termiting this way, by first demonstrating how to do it and then correcting failures by taking away probes made incorrectly or with the wrong materials. However, they do not and the burden falls on the juvenile to pay attention at the right time to the right things. Attention then becomes a key element in primate learning just as it is in the learning of young children.

Attention Structure

We mentioned earlier that one of the aspects of a dominance hierarchy is that it forms what has been called an "attention structure." This term simply refers to the fact that subordinate individuals are usually very much aware of what dominant animals are doing even when the subordinates may seem to be going about their business of feeding, resting and grooming, or traveling. In contrast, very dominant animals do not often pay attention to the behavior of subordinates unless it actively intrudes upon them. Because of this differential in attention, a dominance hierarchy can be seen and analyzed in terms of attention structure—who pays attention to whom. However, the dominance hierarchy is only one kind of attention structure in primate society. There are two other major ones which are unrelated to the dominance hierarchy: the matrifocal subgroup and the peer play group. Members of these two groups also "catch the attention" of each other, not because of fear or awe but because of the special kinds of attractiveness they possess.

A series of very interesting "accidental" experiments in learning have been made by Japanese scientists studying the wild Japanese monkey. In 1952 they began feeding wild groups of monkeys so that they would come daily to a feeding ground where they could be observed. In some areas tourists came to see the monkeys and began feeding them caramel candies wrapped in paper. In another area they had to put out foods which were not part of the regular diet of the monkeys such as sweet potatoes and wheat. The scientists used these changes in the normal diet to study how new ideas and traditions are transmitted through a social group (Itani 1958; Kawai 1965; Kawamura 1959).

The Japanese found interesting differences in the course and speed of transmission of the new food habits according to which animal in the group introduced the new behavior pattern. For example, in 1952 a group of monkeys on Koshima Island were fed sweet potatoes on a small sand beach. Monkeys do not like to eat gritty or sandy food and they normally will brush off dirt before eating. In 1953 a one and one-half year old female developed the habit of taking her sweet potato down to the water and rinsing it off before she ate it. This was in fact a highly adaptive behavior pattern because the habitual eating of gritty food rapidly wears down the teeth, which become broken and abscessed. Although a monkey cannot appreciate the fact that without teeth and with infected jaws it might die, an aversion to eating gritty foods was nevertheless selected for so that the Koshima monkeys could appreciate the change in the substance of a mouthful of sweet potatoes. By 1962 almost all the members of the group had adopted this habit except for a few very old animals.

Kawai and Kawamura kept careful records of the spread of this habit through the group and they found that the trait passed slowly from one animal to another following lines of special relationships, usually ties of affection. The first animal to learn the new trait from the juvenile was her mother. This is understandable, since practically the only adult animal to be interested in the behavior of a one and one-half year old female is her mother. Once the mother had adopted the behavior then it passed naturally to all her subsequent offspring. An older sibling of the juvenile innovator learned the behavior several years later, probably from watching her younger sister. The main line of diffusion of sweet-potato washing was through the play group of the young female. Some of her playmates were interested in her behavior, observed the potato washing, and began doing it themselves. Their mothers and older siblings then learned it and so the habit passed through other geneaologies. Eventually, through the attention structures of the matrifocal units and the play group all but 13 monkeys (mostly adult males) had learned to wash potatoes. The adult males were not resistant to the idea on principle; they simply did not notice what was going on in a way that would affect their behavior.

In the case of eating caramel candies, adult males in another group took up the new food habit when it diffused to them through very special relations which they develop during the birth season. At that time, when monther monkeys are busy with their new infants, adult males of some groups "adopt" recently weaned yearlings for a few months, carrying them about, grooming and sleeping with them, and generally acting like mother substitutes. Some of these males learned to like caramel candy from their little protégés. Once high ranking males learned the new habit, it spread among the other males who were paying attention to them because of the dominance hierarchy. It seemed that in no case do monkeys simply look over, observe a new behavior pattern like sweet-potato washing or eating candies, recognize the activity as useful, and adopt the behavior. The proper emotional climate and interest was only there when there was a positive emotional bond present which made the observer really see what the demonstrating monkey was doing.

Kawai (1965) noticed that the rate of diffusion of a new food habit seemed to depend directly on the social status of the innovator. In one case wheat was given for the first time to a group, only one of whose members was familiar with it. This happened to be a high-ranking male who may have had previous experience with wheat. In this situation the spread was very rapid. From him it went to the dominant male, his peer, and then to the most dominant female. From these individuals it spread independently down the male and female dominance hierarchies, and from each female it passed through her family lines independently. These represent the normal lines of transmittal of group traditions and there is little resistance to this pattern. In a matter of a few hours the new behavior had spread throughout the group.

In contrast, another food behavior, that of wheat sluicing, took years to spread. Wheat sluicing was a very useful invention, by the same female (at that time four years old) who discovered sweet-potato washing. When wheat was given to the monkeys, it was simply thrown out in piles on the beach. Very quickly it would become mixed with sand and the monkeys would have to sit for hours painstakingly picking out the wheat grain by grain. In 1960 the young female began to pick up handfuls and armloads of wheat and sand, dash on two hind legs to the water, and drop them in. The wheat would then float while the sand sank so that she could easily sweep up handfuls of wheat and eat them. To the monkeys sitting on the beach picking up individual grains one by one, this ought to have looked like a very good idea. However, three years later only 14 out of 58 members of the Koshima group had learned to do it. The psychology of adult monkeys is basically conservative; they are not looking for new ways to do things, and when innovation comes from low-status animals it is not appreciated simply because of the attention structure of the group. Although juveniles and subadults are the most experimental and innovative of the age classes, as a subgroup they hold low status so that the end result is that group traditions change very slowly. In the long run this is a very adaptive system which allows for change in group traditions but not so fast that there is instability and disorganization.

As van Lawick-Goodall (1970) pointed out, the contribution of individuals to the traditional behavior patterns of their groups should never be underestimated when dealing with the higher primates. There are certain individuals, such as the young female Japanese macaque who invented sweet-potato washing and wheat sluicing, and Mike, the chimpanzee who learned to use kerosene cans for his display, who are able to invent new behaviors or to put together old behavior patterns into wholly new combinations (Fig. 3–1). Over a lifetime the presence of such individuals can greatly influence the behavioral repertoire of a group and its adaptations to the environment.

Observational Learning

The work done by the Japanese strongly suggests that young animals are extremely interested in the behavior of others about them—their mothers, other line-

Figure 3–1. The adaptive value of group traditions. Japanese monkeys pass a cold winter's day grooming and relaxing in a hot spring. The behavior pattern, a group tradition now, began as play among juveniles.

age members, and peers. They watch intently the feeding behavior of others, even before they are weaned, and they begin playing with food long before they have to feed themselves. When new food such as caramel candies are introduced, they watch intently to see what the mother's reaction will be. Even if she is indifferent, the infant will probably investigate the food anyway and after a series of sniffs, nibbles, and inspections, begin to eat it. Around the age of three and one-half, which is puberty for these monkeys, a conservatism begins to develop which is firmly established by the age of eight. Curious investigation and observation gradually stop so that fully mature animals rarely learn new foods, with the exception of some mothers who learn from their infants. The Japanese found that almost no males over the age of four were able to learn new patterns or food items whereas juvenile males learn just as readily as juvenile females.

Van Lawick-Goodall has had the opportunity of seeing how young chimpanzees learn complex behavior patterns such as termiting. One young male about seven months old began "mopping up" movements about two weeks after the termiting season began. "Mopping up" is van Lawick-Goodall's term for the way chimpanzees use the hair on the back of their wrists to pick up termites which have accidentally fallen to the ground from their probes. The infant male began mopping things indiscriminately—the ground, trees, his mother. He also started to pick pieces of grass and poke at the termites on his mother's arm.

Infant chimpanzees only play at termiting during their first year but by the time of their second rainy season, the play becomes more organized and focused. One young female began at the age of one and one-half to make her own termiting tools. At first her technique was very imperfect. She made tools that were too short to more than just enter the hole; the longest was only two inches whereas the adults always use six to twelve inch straws. The infant's motor patterns were imperfectly coordinated and sometimes she would jerk the straw out of the hole so quickly that the termites were knocked off. Her attention span was very short as well; she would termite for a few minutes and then break off to play. In contrast, adults often work with great concentration for more than an hour without stopping (the adult record was five hours for a mature female). Young animals were also seen trying to termite out of season in what may have been a form of play activity.

Van Lawick-Goodall is convinced that much of the ability to termite is learned by young animals by first observing the adult technique and then practicing it. She often saw an infant intently watching an adult termiting and then, when the adult had moved off leaving the straw or twig by the nest, the infant would pick up the abandoned tool and try termiting too. The Japanese noted the same process among yearlings in the Koshima group. At first they only watched their mothers very closely as they gathered up handfuls of wheat to throw in the water. Then the infants would go through a period when they would scratch at the sand as if they were trying to imitate the behavior without understanding its purpose. Only later would the full behavior pattern develop, after much practice. Of course, this is very similar to the learning of both social and motor skills in human children, elements of the pattern are imitated long before they are understood or integrated into any useful behavior.

Play

PLAY

Although observational learning is very important in terms of the evolution of social traditions, the most significant aspect of learning in primates is play. All field workers have been impressed with the amount of time and energy spent every day in play by juvenile primates. For most species this means social play hours spent in active chasing and wrestling, and playing such pan-primate games as hide-and-seek and king-of-the-mountain. Play, like other categories of behavior, varies from species to species according to their adaptations. For example, vervet monkeys in Zambia have four major kinds of play behavior. Two of these relate to their predator–prey relationships and two to social relations within the group.

Vervet monkeys spend much of their time on the ground during the day in spite of the fact that they are relatively small (fully adult males average 15 pounds, adult females 9 pounds). Their small size, and the fact that they sometimes feed in the open away from trees or in tall grasses, means that they are potential prey to many species, even eagles and owls. It is reasonable to interpret many of the play activities of juveniles as practice for escape from predators. One of the most popular types of play is hiding and chasing in tall grass. The usual form it takes is for a group of juveniles to space themselves out in a grassy area. Suddenly, as if on signal, they will all disappear into the grass. The first one to move or make a noise will be jumped on by all the others. This kind of play is practice for the way vervets avoid predators when they are not close enough to run to the safety of trees: they hide in the grass without moving and depend on the predator not seeing them as it passes.

Another common form of play among vervets is circular tag in which a group of juveniles will chase each other up into a tree, out on a branch, jump to the ground, and then run back up in the tree. As Dolhinow and Bishop (1970) have noted, young monkeys learn from this in two ways. Through constant repetition they will learn the physical properties of their environment and how to adapt to them, such as how bouncy a long limb might be and how to adjust their leaps accordingly. Secondly, the enjoyment of play provides the motivation for the kind of daily exercise and practice of physical skills that is necessary for survival in an environment where speed and coordination can mean safety from predators. Being at top form might well spell the difference between capture and escape for a species which is too small to fight for its safety.

Another form of play in young vervets is one which combines various elements of common adult behavior patterns. These elements are drawn from behavioral systems such as fighting, mating, and grooming. In play these patterns become mixed and roles are changed rapidly back and forth. A pair of monkeys might begin by clasping each other, wrestling and tumbling about until one monkey breaks away and is chased by the other. In the middle of the chase the fleeing monkey may suddenly turn and begin chasing his pursuer. Mounting with pelvic thrusts and grooming may also get mixed in with the play-fighting, again with roles changing back and forth. Although this kind of play may not seem to be practice for adult life because of its mixed character, in fact it gives juveniles physical experience in performing behavior patterns in the adult repertoire without the usual social results or consequences. Sometimes, of course, play-fighting gets out of hand, just as with human children, and somebody gets hurt and squeals. This usually puts an end to the play for the moment because the squeal draws the attention of the adults. One of the characteristics of monkey social play is its complete silence as long as everything is going along well. Van Lawick-Goodall reports that sometimes she can hear soft chuckling when chimpanzees play. It is only human children, who have little fear of predators, whose play is filled with noisy yelling and screaming.

The final form of play in vervets is play-mothering. This behavior is not unique to vervets but they do represent an extreme example. Mother vervets permit juvenile females to take and handle, groom, and carry infants within a few weeks of birth (Lancaster 1971). The beginning of the birth season each year marks an abrupt change in the behavior of juvenile females aged one to three years. Before the infants are born, juvenile females spend most of their time either in the company of their mothers or in play groups. However, the presence of a newborn infant in the group acts as a magnet to a juvenile female. It is common to see each new mother acquire an entourage of juvenile females who follow her about during the day waiting for a chance to touch or to hold the infant. Often a juvenile female will begin by grooming the mother, only gradually, but very obviously, to the observer, working her way over toward the nursing infant. If the mother moves or seems disturbed by this attention to the infant, the juvenile female will hurriedly begin to groom the mother again. During the early weeks the young female will try to gently pull the infant away from its mother and hold it to her own chest or groom it. By the time the infant is three weeks old the mother is usually much more relaxed about it, and she may let juvenile females hug and carry it for several minutes.

When very young females get their first chance to hold an infant or carry it, they often have difficulty in orienting its body properly. Sometimes the infant refuses to cling, and a juvenile female will clutch the infant to her chest with both arms and run bipedally to try to get it out of its mother's view. Females soon learn how to carry infants properly and they also learn that, if they can keep an infant quiet and content, its mother will probably not try to retrieve it. She learns to groom it to keep it relaxed, and often can be seen pinning down an infant with a leg or arm and then intensely grooming it until it stops struggling. By the end of the birth season young females clearly have developed much skill in handling infants. The awkward movements of the first weeks are replaced by casual, skilled maternal behavior with which infants can be scooped up onto the chest in an emergency or carefully and comfortably groomed. Undoubtedly this skill and familiarity with infants is useful for the young female when she matures and has her own first baby. The behavior is probably beneficial for the infants, too. They get extra attention from a segment of the social group with nothing more important to do. In a dangerous environment and at a time when the mother must do extra feeding to maintain her milk supply, the help in baby-sitting must have real adaptive advantages to both the mother and her infant.

Although play and observational learning may differ according to the species, they have one important common characteristic. It is only the carefree mind that is free to learn new behaviors. Observational learning and play behavior are inhibited by fighting, fear, anxiety, or anything else that creates tension. Among vervets as well as baboons and macaques, a common scene is a play group fully abandoned in its wild chases while an adult male sits vigilant nearby. It is the social system which provides the kind of emotional atmosphere which is prerequisite to the learning of social traditions.

SOCIAL TRADITIONS AND THE EVOLUTION OF CULTURE

Chimpanzee Tool Use

The chimpanzees of the Gombe Stream Reserve in Tanzania are probably the best studied of any wild population of primates. Van Lawick-Goodall began her work in 1960 and since that time the local population of about 60 chimpanzees has been under observation by her and a number of ethologists, zoologists, anthropologists, and psychologists who have joined her for various periods of time. This remarkable long-term study has yielded vast amounts of information on the chimpanzee social organization, family life, ecological adaptation, and communication system. Chimpanzees are very closely related to human beings according to a great many measures: anatomy of the body, serum proteins, chromosome number and form, and dentition. The field studies have added to this list, establishing many behavioral continuities between humans and chimpanzees in such things as mother–infant relations, gestures and sounds used in communication, meat sharing and cooperative hunting, flexibility of the social group, and long-term social relationships among adults. As remarkable as these observations have been on social behavior, the observations on tool using have been even more so. There can be little doubt now that tool using by chimpanzees excels that reported for all other animals except human beings in both variety and complexity. Furthermore, chimpanzees not only make and use a wide variety of tools, but this behavior seems to represent social traditions of the group passed on from one generation to the next.

Chimpanzees use tools in two very different contexts (van Lawick-Goodall 1970): first when the individual is feeding or doing some other nonemotional activity and the second in the context of excited, aggressive display. Termiting is only one example of tool using by chimpanzees as an aid to feeding. They have been seen using straw or stick probes when feeding on ants and to dip into bees' nests for honey. They will also use sticks as probes or levers to try to pry something open or up. At the Gombe Research Center chimpanzees tried to pry open the lids of boxes of bananas with sticks. They would first break off a suitable stick, then strip it of leaves, and often bite splinters off one end so that it formed a chisel-shaped edge. In other parts of Africa stones used by chimpanzees to bash open hard-shelled fruits have been found (Struhsaker and Hunkeler 1971).

Chimpanzees also make leaf sponges for dipping water out of crevices and boles of trees that are too small to let them put their faces down to the water. They take a handful of leaves, chew them slightly, dip the wad into the water, and then suck it. Van Lawick-Goodall tried the same thing and found it seven or eight times more efficient than the technique used by many primates of dipping the hand or fingers into water and letting it drip into the mouth. Besides using leaves as sponges for drinking water, chimpanzees use them to wipe water or dirt from the body, dab at wounds, or remove sticky substances from the fingers. They also use sticks and twigs to pick their teeth. In captivity one chimpanzee may groom another's gums and teeth using a stick (McGrew and Tutin 1973). Chimpanzees with diarrhea may wipe themselves clean with a large handful of leaves and twice van Lawick-Goodall observed a three year old, after intently watching his mother wipe her bottom, pick leaves and do exactly the same to himself.

The other way in which chimpanzees use objects as tools is in aggressive display. This behavior is particularly interesting because it suggests that tools for defense may have developed just as early in human history as tools for food-getting. Random throwing of objects—anything that comes to hand such as stones, sticks, or other vegetation—is a common element in the excited displays of many monkeys, apes, and humans, and chimpanzees are no exception. They tend to throw things when meeting other groups after a separation or when being annoyed by baboons or humans. Sometimes a chimpanzee will even take some care in aiming the object; instead of just tossing it into the air, he will throw it toward the animal at which the display is aimed. Van Lawick-Goodall saw chimpanzees aim and throw stones, both overhand and underhand, at baboons and at humans as part of such a sequence of aggressive display. This behavior pattern is significant because, as Washburn (Washburn and Moore 1974) has pointed out, it suggests the possible first step in the evolution of weapons. If an animal is displaying to intimidate an aggressor, object throwing as a part of that display is effective whether he hits the other animal or not. If the total display is not intimidating enough, the chimpanzee is still able to flee or to fight with his canines. Van Lawick-Goodall reports that in fact chimpanzees only use objects in display and not in true fighting. As soon as actual physical fighting is imminent, they will drop the stick or stone and attack with the hands and teeth. It is easy to imagine how the ability to develop skill in aimed throwing of sticks and rocks could gradually evolve until it became so effective that the creature need no longer rely on his canines. Only then would the selective pressure on large canines be relaxed and a new approach, defense with weapons, ultimately replace the behavioral and anatomical pattern of defense by fighting with canines.

These examples of tool use in chimpanzees, when taken together, provide a good starting place for answering questions about how and why tools were used by the human species' earliest ancestors. It is true that some birds and other mammals use tools, but in any one tool-using species there is likely to be only one kind of tool employed. There is no nonprimate that uses such different objects as termiting straws, leaf sponges, and stone projectiles. And, conversely, in the chimpanzee there is no single, highly evolved, stereotyped sequence of movements of the sort common in other tool users, such as the deft twist of a cactus spine used by finches to dig grubs from bark. In the chimpanzee there is a far more generalized tendency to manipulate objects and to use them in many different situations. This single population of chimpanzees at Gombe has performed more complex kinds of tool use, and in a wider variety of situations, than has been observed for any other animal; and furthermore, that is true even though tool use is a very small part of their behavioral repertoire and is a comparatively rare event.

The tool-using behavior of chimpanzees suggests the kind of ape ancestor that might be postulated for the origin of the human line—an ape that used tools for many different reasons and in many different ways, no matter how insignificant the tool, like termiting straws, or inefficient, like a clumsily swung stick. The more kinds of implements this ape used the more likely his ancestral role, because it would have been the accumulated influence of many ways and reasons for using tools that would have taken selective pressure off the specific situation, tool, and movement. Selective pressure favored a hand that could use many tools skillfully and a brain capable of learning these skills. Natural selection would then have acted upon a broader category of behavior, one involving the brain, the hand, many objects, and a wide variety of social and ecological situations and problems. The evolution of skilled tool using marks a major change from the kind of tool use that is incidental to the life of a chimpanzee to the kind that is absolutely essential for survival of the human individual.

Human Skill and Culture

Wild chimpanzees have proven to be far more advanced in tool making and tool using than most people really expected, in spite of their cleverness in zoos and the laboratory. Their natural behavior displays many of the prerogatives for the evolution of a human, tool-using culture. Once given the possibility of techniques of use and manufacture being learned by one animal from another, in combination with an open system which encouraged all kinds of tool using, the evolution of a species committed to tools became almost inevitable.

The lack of divergence in anatomy and behavior between humans and the African apes strongly supports the idea of an end of the Miocene or early Pliocene (but no earlier) division between them, somewhere on the order of 10 million years ago. If this is so, then it may be worth considering the possibility that casual, unskilled tool use might have been typical of many species of apes during the Pliocene. It should be remembered that the late Miocene and early Pliocene represent a time when apes were abundant, diverse, and widely spread over much of the Old World (Europe, Africa, India, and China). The modern apes are only remnants, survivors of a time when the family was highly successful and diverse. Table 3–1 suggests that a kind of

TABLE 3–1   CHRONOLOGY OF THE EVOLUTION OF TOOL-USING BEHAVIOR

Geological Time Divisions	Radiometric Age Estimates	Tool-using Species of Primates	Types of Tool-using Traditions
Pleistocene	1,000,000 B.P.	Homo erectus (1 species only)	Hand axe and later traditions
	4,500,000 B.P.	Australopithecus (more than 1 species)	Pebble-tool industries
Pliocene		Ape and Human lines (many species)	Unskilled, ape tool use (hypothetical)
	10,000,000+ B.P.		

unskilled ape tool use continued down to the present in one or more of the surviving, descendant species of ape (Lancaster 1968a).

The beginning of the Pleistocene witnessed the emergence of perhaps one or more forms of bipedal human, the Australopithecines who, although possessing relatively small brains, had come to rely on tool use for much of their food getting and defense. Clearly specializations in the hands and especially the thumb, the small size of the canines and incisors, and bipedalism all point toward the importance of tools in their way of life. Both the small brains and the tools themselves suggest a lack of skill in the way the implements were made and used (Washburn 1968; Washburn and Lancaster 1968).

The key to understanding the evolution of skill probably lies in the evolution of handedness. Chimpanzees are ambidextrous in their tool using; all animals can use either hand although there may be some individual preference for one over the other. This absence of handedness may be indicative of the limitations that the chimpanzee's brain places on its ability to develop highly skilled tool use. By human standards the movements used in termiting and other kinds of tool-using behavior always appear clumsy, like the use of implements by a human child. Tool use is learned by the chimpanzee and improves with practice, but never develops the deftness either of human skill or of highly stereotyped, innate motor patterns.

The archeological record of human tools suggests that from the time of the earliest known stone tools, about 3 million years ago, until about 1 million years ago, there was almost no change in the techniques of tool making. The basic stone tool kit included rough chopping tools, crudely trimmed flakes for knives and scrapers, and round bashers. There seemed to be no special standardization in tool-making techniques and no regional specialization of tool kits. Then, about a million years ago, a rapid rate of evolutionary development in brain size and complexity of tool assemblages seems to have begun. This later period is associated with the emergence of a single species of tool user, *Homo erectus*, dominating much of the Old World. Remnant species of ape also survived but with quite restricted geographic distribution. Perhaps a new efficiency in the skilled use of tools effectively closed the niche to competition. It probably left no room within the broad niche created by tools for separate species to develop specialized applications. Any possibility for different kinds of tool users—perhaps an open savannah, a woodland, and a forest form (a possibility that may have been realized in the Australopithecines of the early Pleistocene)—disappeared and a single species of human, using various tool traditions, spread across the Old World. The increase of efficiency and skill with which tools were used, a trend that probably began in the early Pliocene, may well have been associated with a gradual decrease in the number of primate species able to command a portion of the niche open to tool users. At an early, inefficient stage many species may have tried using tools with variable degrees of success but, as skill and efficiency developed, the competition between tool-using species increased, and the possibility of many forms sharing the niche disappeared.

TOPICS FOR DISCUSSION OR PROJECTS

1. Considering what you know about learning in nonhuman primates, how would you design an ideal school for young children? Discuss such aspects as curriculum, teaching techniques, setting, group size and composition, relationship with the teacher.

4

Primate Communication and the Emergence of Language

INTRODUCTION

The interest in human evolution and in the origin of human language has distorted the study of the communication systems of the nonhuman primate. These systems are not steps toward language. The cries, displays, and facial expressions given by both human and nonhuman primates in social interaction are largely expressive. They communicate information about the motivational state of the signaller. The mutual exchange of such signals between monkeys is not comparable to human language but has much more in common with the signals used in social interaction by many birds and mammals. Human language is unusual in its ability not only to express emotion but also to make reference to the environment. This unique evolution of a referential function in communication rests upon special changes in the brain during human evolution. The purpose of this chapter is to examine this major gap that separates language from nonhuman primate communication and to suggest conditions that might have led to the evolution of language.

Human beings are so accustomed to many sounds having definite, restricted meanings that the first question asked on hearing a monkey sound is often, "What does it mean?" In our experience it takes some time for a student to lose this human bias and to understand the sounds in the context of normal monkey behavior. Humans expect sounds alone to carry meaning, but in nonhuman communication they usually carry only a part of the message, and facial expressions, gestures, and postures are essential in conveying the full meaning. The sound may be only for emphasis and, even in the case of warning cries, the warning may give no indication of the exact nature of the danger. In contrast, human language is essentially a system of names, governed by grammatical rules, and most of the meaning of the communication is carried by a single sensory mode, hearing.

Most acts of communication in a social group of primates occur in a context of long-term social relations, since monkey and ape societies are usually composed of animals of both sexes and all ages and most members of the group have spent their entire lives within the same social structure. Even in species where there may be no encompassing stable group there are still stable subgroups with continuing, long-term relationships. Communication rarely occurs between true strangers but for the most part between animals that have known each other as individuals over long periods of time. The context, then, of any communicative act includes a network of social relations that have a considerable history behind them, all of which is relevant to the message and how it is received and responded to.

Let us begin by comparing the way in which a chimpanzee group might communicate about a particular situation compared to the way humans might. Let us say that both groups are wandering through a stretch of woodland. The group members are spread out as they forage—often only a few are in visual contact with each other. A young adult male comes upon a small tree with ripe fruit, enough for each member of the group to have only one or two pieces. If the young male is a chimpanzee, the amount of information that he can transmit to comrades is severely limited by his communication system. He will vocalize—giving a series of hoots. These hoots will indicate to others that the young male is pleasantly excited. The intensity, duration, and frequency of the hooting will suggest exactly how excited the male is. The direction of the sound will help others find him. The communication system gives the chimpanzee no way to express what he has found, where it is, how much there is, or what he would like others to do. In contrast, a human in the same situation can give a message such as: "I have found a tree with ripe figs by the river. I need two people to help me gather them. We will catch up with the rest of the group later at the waterhole." The human, too, will convey urgency and excitement through the loudness and tone of his voice and its direction will help others find him, but language enabled him to convey far more environmental information, which greatly increased the efficiency with which the group as a whole exploited its home range.

An understanding of the emergence of human language rests upon a comprehension of the factors that led to the evolution of a system of names to refer to the environment. It is true that a dog can be conditioned to form an association between an object in the environment and his own internal state. In this way a dog may be trained to bring his bowl to his master when he is hungry. Similarly, a chimpanzee was trained to make a vocal sound something like "cup" to indicate she was thirsty. But these conditioned associations between an environmental stimulus and an internal state are not really steps towards language, and furthermore, there is probably a limitation imposed by the brain on how many such associations a dog or chimpanzee can be taught. In contrast, human discourse is based on a practically unlimited ability to form associations between two different sets of environmental stimuli: the sound of a word and the sensory image representing the environmental referent of that word. There have been special changes in the human brain during our evolution which allow us to form many such associations and to form them with great ease. The ability to use names allows humans to refer to the environment and to communicate information about it as opposed to the ability to express only our own emotions. Naming is the simplest form of environmental reference. It is an ability that is developed to a unique degree in human beings. In itself it is not language, but without it human language cannot exist. It is important to understand the communication systems of nonhuman primates if human language is to be understood, not because they are similar, but because they are different; understanding in this case comes from analyzing contrasts, not similarities.

STUDIES OF PRIMATE COMMUNICATION

The study of the communication systems of Old World monkeys and apes is a new and rapidly expanding field. Only a few workers have as yet been able to focus on the system itself—recording a large sample of sounds, gestures, and expressions and struggling seriously with the infinite variety of problems of description and analysis. A number of studies have now been made on colonies living in compounds, on free-ranging groups that are artificially fed, and on free-ranging groups living in their natural habitats. However, most studies on primate communication have been on either baboons or macaques—two closely related genera of ground-living Old World monkeys which strongly bias the information we now have on primates. A second sampling bias has also developed because of the comparative ease of recording and describing sounds through the use of electronic recording equipment and the sound spectrograph. Although photographic equipment is easily used, there is no photographic equivalent of the sound spectrograph when it comes to describing and analyzing complex patterns of movement found in facial expressions and gestures. The result of these two biases in sampling is that we know a fair amount about the nature of the vocalizations of two genera of Old World monkeys, and much less about the communication systems of most of the Old World monkeys and apes.

THE FORM AND NATURE OF SIGNALS IN
PRIMATE COMMUNICATION

One of the most significant generalizations which can be drawn from primate field studies is that the communication systems of monkeys and apes are extraordinarily complex compared to that of many birds and primitive mammals, and that higher primates rely heavily on multimodal signals. "Multimodal" means that all the senses may be brought into play in receiving a single message. A female rhesus macaque signals her sexual receptivity to a male by approaching him directly with a confident walk. She will make smacking sounds with her tongue and lips, and then either stop before him and begin to groom him or turn and present her rear. In this example, she will make use of elements from all four sensory modalities: tactile (her touch as she grooms him), visual (her posture and walk, presenting her rear), auditory (lip smacking), and olfactory (the odor of her perineum and vagina during estrus). In fact, a vocalization, gesture, or facial expression in itself usually does not represent a complete message but is only a part of a complex constellation of sound, posture, movement, and facial expression. Parts of such a complex pattern may vary independently and may help to express changes in intensity or level of motivation. Sometimes olfactory elements are also present in the signal pattern, but in monkeys and apes and in humans the senses of sight, hearing, and touch are much more important in receiving communicative signals.

One obvious reason for the large number of multimodal signal patterns can be found in the kinds of social groupings typical of monkeys and apes. Many species of

Old World monkeys and apes live in groups of perhaps 15 to 30 animals. These groups tend to be cohesive, and individuals are often spatially very close to each other. This kind of social life permits the evolutionary development of multimodal signals because social situations are constantly recurring in which tactile, auditory, visual, and olfactory signals can be sent simultaneously. When signals have to pass over greater distances, touch and smell are useless to higher primates. Sight and sound become more significant, and vocalizations, which are often comparatively unimportant in close-range systems, may carry the major burden of communication when long distances need to be covered. Many field and laboratory workers have emphasized that vocalizations are not particularly important in most primate social interactions but function instead either to call visual attention to the signaler or to emphasize or enhance the effect of visual and tactile signals. In other words, a blind monkey might be greatly handicapped in his social interactions whereas a deaf one would probably be able to function almost normally.

Another important generalization that has emerged from the field studies of monkeys and apes is that, just as with other animals, context plays a major role in the total meaning of the signal pattern. The receiver of a signal is presented with an extremely complex pattern of stimuli: posture, gesture, vocalization, and facial expression of the signaling animal are important, but the total context of that pattern is also an essential part of the message. The immediately preceding events, the social context, and the environmental context all play major roles in the way a signal is received, interpreted, and responded to. A threat display given by a juvenile may be ignored in one situation, whereas if the same display is given again when he is near his mother, and especially if she shows some interest in what he is doing, it may produce an entirely different response in the animal receiving the threat. The major function of context in the total impact of the signal makes the study of primate communication systems very difficult. Responses to a signal pattern may seem highly variable and erratic until a large number has been sampled and the relevant aspects of the varying contexts of the signal have been taken into account.

Besides being multimodal, primate signals are often graded in form, that is, variations that may reflect differences in meaning occur in a single behavior pattern, such as in a threat gesture or vocalization. In a graded or continuous system of behavior patterns, each grade or degree has at least the potential for expressing slight differences in intensity of motivation. The advantage of discrete, nongraded signals, of the sort typical of many birds, is that their lack of ambiguity makes them easy to receive and to comprehend. A communication system based on signal patterns both limited in number and stereotyped simply requires less learning and less cerebral cortex than does one based on many unstereotyped, graded signals which place greater demands on the receiver of the signal, but which also have great value in their ability to express slight shifts in motivation. In a complex, enduring social system in which individuals are obliged to make a continuous series of adjustments and accommodations to each other, it is important to be able to express not just that one is aroused or frightened but also the degree and direction of changes in motivation.

A good documentation of graded signals is found in Hinde and Rowell's description of communication in a colony of rhesus macaques (Hinde and Rowell 1962;

Rowell 1962; Rowell and Hinde 1962). In this system many signals were not only graded in form but also intergraded with each other. Rowell and Hinde demonstrated this by making spectographs of all sounds that occurred in agonistic situations. They found what they thought were nine harsh sounds ranging from a growl to a squeal. After a large number of these had been recorded and analyzed with a sound spectograph, they discovered that the sounds in fact formed a single intergrading system that seemed to be expressive of the full range of emotion usually associated with agonistic interactions. These agonistic sounds were linked by a continuous series of intermediate forms, and apparently each grade along the continuum potentially expressed a slightly different level of emotion (Fig. 4–1). There was also one example of a multidimensional variation in which the pant-threat graded independently into three other calls—the roar, the bark, and the growl. With such a system, a rhesus monkey is able to express quite complex patterns of motivation, but most of the variations in signal form rest on contrasts in intensity of one or more of the most dominant components of the total motivational state. In concert, this use of intergrading signals and of composites from several sensory modes produces a rich potential for the expression of very slight but significant changes in the intensity and nature of the mood of the signaling animal. Slight shifts or vacillations in arousal can be expressed by slight changes in the vocalization and gestures.

Not all primate signals belong to graded systems, and there are undoubtedly species differences in how much use is made of discrete or graded signals. Struhsaker (1967) has described the vocalizations of vervet monkeys which he recorded in their natural habitat. He found 36 different sounds that were comparatively distinct both to the human ear and when analyzed by a sound spectograph. The majority of vervet sounds seem to be of the discrete type, although there were two groups of sounds that may form graded systems. With more and more study on primates it will probably be shown that their communication systems tend to be of a mixed type in which both graded and discrete signal patterns are used depending on the relative efficiency of one or the other form in serving a specific function. In such systems of communication as those of the monkeys and apes, complexity and subtlety of expression are bought at a sacrifice to clarity and specificity. With complexity comes ambiguity, and greater burdens of discrimination and interpretation are placed on the nervous system of the receiver of the signal, which in turn places potential limits on the communication system.

THE NATURE OF THE MESSAGES AND THEIR RELATION TO THE LIMBIC SYSTEM

It is clear that the communication systems of monkeys and apes are rich in their ability to express the motivational state of the animals. Such messages facilitate social interactions. In baboons and macaques, motivational information—particularly in relation to dominance and subordinance relationships—constitutes the largest category of messages. Even greetings and other messages exchanged when one animal approaches another often serve to reassert recognized differences in

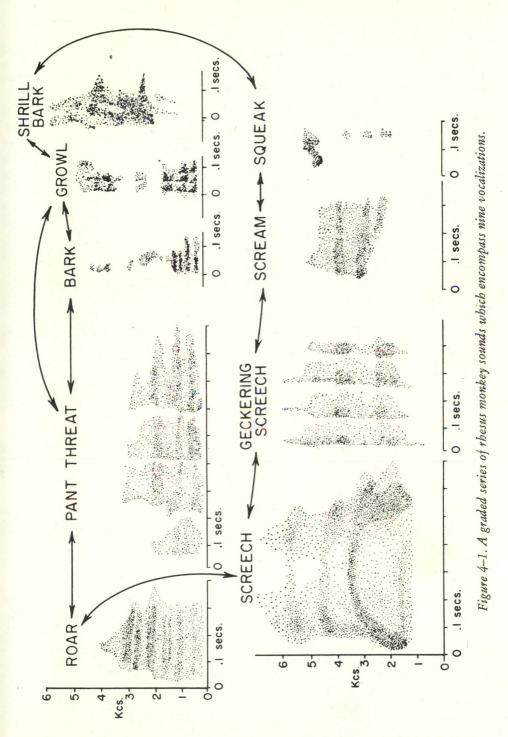

Figure 4–1. A graded series of rhesus monkey sounds which encompass nine vocalizations.

dominance between two animals. Sometimes the dominant animal will gesture or indicate in some other way the pacific nature of his intentions at the same time that he displays his dominance. Compared to birds or more primitive mammals such as hamsters, monkeys and apes have developed very complex signal patterns expressing submission, aggression, anxiety, fear, and other motivational states associated with agonistic situations. There is also another large category of signals which are mainly found in friendly, bond-maintaining, or bond-developing situations such as play, mating, and mother–infant interactions, as well as signals that keep the group together and coordinate group movement. It is interesting that in nonhuman primates, just as in man, many of these signals seem to derive directly from very early mother–infant interactions. That is why we find adult monkeys, apes, and human beings hugging, stroking, patting, embracing, kissing, and nuzzling others when they feel affectionate. Similarly, many adult primates when distressed and seeking help will give cries, whimpers, or cling to others in a way that is reminiscent of infants.

The limitations of such a signal system lie in the fact that it really functions to express only a few motivational states such as aggression, desire for physical contact, or high excitement. These states are closely related to the functions of the limbic system, a part of the brain which has a long evolutionary history. The limbic system is found as a common denominator in the brains of all mammals, and even with the expansion of the cerebral cortex during human evolution, it still remains a major part of the human brain. Stimulation of a part of the limbic system usually elicits not only behavioral patterns such as the movements used in aggression or mating, but also emotions that provide the animal with a state of heightened motivation to perform a behavioral pattern that is critical to its survival (Hamburg 1963). The limbic system is closely associated with emotion—it makes the animal want to do what it has to do to survive and reproduce.

In a system communicating only internal motivational states, what can be communicated is limited by the number of states, modified only by their intensity. Thus, variations within each basic sound class—an aggressive bark, for example—are largely related to variations in intensity of the particular emotional state. Transitional and combined calls may often represent motivational states in which more than one emotion may be aroused. Winter, Ploog, and Latta (1966) found that the squirrel monkey's squawk is structurally a combination of a shriek (high excitement) and a twit call (distance promoting). It is often given when one squirrel monkey steals food from another; it can be given by either animal and it probably expresses high excitement and a desire to maintain distance between the two actors, both of which are emotions appropriate to the situation.

The relationship between a primate sound or gesture and the underlying motivation which it expresses can be made clearer by comparing it with language. In human language the particular set of sounds that are used to indicate a specific empirical phenomenon are largely arbitrary. One human group may use a word to express "fear" completely different from the words used by other human groups to refer to the same phenomenon. In nonhuman primates and in many of the human species' expressive gestures this relationship between a sound or other signal and the emotion it expresses is more closely fixed. It is part of the behavioral repertoire of the species and not just of the local group. For example, the basic set of gestures and sounds described for the rhesus macaque in the laboratory by Hinde and Rowell is the same as that described for a colony of rhesus monkeys living for over 25 years on Cayo Santiago in the Caribbean, and for rhesus monkeys living in their native habitat of North India. Variation is mainly at the quantitative level, so that intergroup differences may occur in the relative frequency of use of certain sound categories. Sometimes sounds which are heard in the natural habitat—such as snake predator calls—may never be heard in the laboratory because of its protected environment.

This fixed relationship is modified by learning only in certain directions. Cries, displays, and other forms of expressive signals in primates are particularly insensitive to training (Skinner 1966). As Skinner has noted, vocal responses of some animals which resemble instinctive cries have been conditioned in the laboratory, but much less easily than responses using other parts of the skeletal nervous system. Primate communication systems do rely on learning and they form part of the traditional patterns of behavior passed from generation to generation. However, it is learning in terms of variation on a theme which belongs to the biology of the species. The basic form of the displays and their general relations to arousal level or emotional state are almost certainly unlearned. Mason (1960, 1961a, 1961b), in controlled laboratory studies comparing the behavior of rhesus monkeys raised in the wild with those raised in the laboratory, found the basic rhesus repertoire of communicative gestures, postures, and vocalizations present in both groups. However, in the laboratory-raised monkeys the use of these signals in a social context lacked the refinements and subtleties of expression necessary for smoothly running social interactions. Orderly social interaction requires reciprocity between sender and receiver; both must be sophisticated in the use of the communication system. The wild-raised animals almost never had to resort to physical coercion like biting or hitting to gain their way, but instead used less direct threats such as lunges in place, barks, and fixed stares. The lack of sophistication with which the laboratory-raised animals used the communication system meant that they often had to resort to maximal threats and fighting. Learning, then, can occur, but it does not radically alter the phylogenetically established relationship between the display pattern, the motivational state it expresses, and its function in the social system. Learning in this case is a process of refining patterns which belong to the behavior repertoire of the species.

The evolution of a species repertoire to communicate internal motivational states, that is emotional "displays" as they are called by students of animal behavior, depends on how effectively each display mirrors the underlying motivation. If a large, dominant male threatens strongly to a subordinate adult who then retreats from a contested food source, both animals benefit. The dominant male has not had to fight for the piece of food even though he was ready to do so. Although he was clearly stronger than the other animal, it was still better for him to avoid the fight and not risk receiving a wound which could become infected and ultimately kill him. As for the subordinate animal, he was able to assess how strongly the dominant animal felt about having the food to himself. Perhaps the subordinate would have responded to a halfhearted threat by slipping up and stealing a bit of the food, but the strong threat told him that going for the food was risking a fight. In such a system, if the outcome of a fight is predictable, then both parties benefit from not fighting. However, the predictability of the outcome of the fight comes from the effectiveness of the communication system. There is no place for lying in this system—perhaps for slight exaggerations of the truth, but nothing more.

Nonhuman primates can send complex messages about their motivational states; they communicate almost nothing about the state of their physical environments. Human beings possess a communication system that is highly evolved in its ability to make environmental references, but this is a distinctively human specialization that should not be taken for granted in monkeys and apes. It may not even be appropriate to assume the existence of such a simple reference to the environment as a food call. It is probable that what have been labeled food calls by many field workers are really expressions of a general level of excitement, which is often associated with food but which may be given in other circumstances as well. For example, Andrew (1962) reports that in many species of primate the same sounds that are given at the sight of food are also given in greeting a companion. In both instances the animal vocalizes upon perceiving a desired object; it is not giving a food call in the sense of making reference to specific items in its environment.

Even in a call warning of a threat of predation on the group, not much specific information about the danger itself is necessarily given. With baboons, when a low- or medium-intensity alarm cry is made, the other animals turn first to see at what the calling animal is looking (Hall and DeVore 1965). The cry itself gives no specific information about the nature or location of the danger, but only indicates the level of excitement or alarm of the animal that first gave the call. Only by looking does the rest of the group learn what is the cause of the state of alarm in one of its members. A baboon responds to the alarm cries of birds and other mammals just as readily as he does to those given by a member of his own social group.

There are a few calls given by nonhuman primates that convey some information about the physical environment. They are rare and represent important specializations of the few species that use them. For example, in situations where a monkey species is regularly preyed upon by different kinds of predators an elaboration of alarm cries may occur. Struhsaker (1967) describes three high-intensity alarm cries of vervet monkeys that are very different in form and that evoke very different responses: a snake chutter, another call given when an airborne predator is seen, and a chirp that signals a terrestrial predator. There is an appropriate and different response to each of these calls. The snake call evokes a mobbing response similar to owl mobbing by birds. On hearing the call signaling an airborne predator, vervets seek cover either by running into tall grass or by dropping out of the tree branches into the dense thickets below, depending on where they were when they first heard the alarm. The response to the chirp that warns of a terrestrial predator is exactly the opposite of that elicited by the call that warns of an aerial one: the vervets run to the trees and go out onto the ends of branches, which would be dangerous if the predator were airborne (a monkey-eating eagle, for example), but which is safe when a predator such as a lion is on the ground. This kind of specialization in vervets, in which some limited but vital information about the environment is communicated, has evolved in many different species of animals, ranging from chickens to rodents, and can be expected when a species is hard pressed by such different kinds of predators as snakes, birds, and large mammals. This differentiation of high-intensity alarm calls to communicate some information about the environment is a specialization that should not be thought of as pointing toward the kind of major revolution in information content suggested as a requisite of human language.

One of the limitations of a communication system which is highly evolved to express emotional states, but which cannot easily refer to the environment, is illustrated by an example of learning processes in monkeys. As was explained earlier, mother monkeys and apes do not purposively teach their infants anything. There is ample evidence that the infant learns many behaviors from its mother and other adults in the group, but all of this learning occurs without active instruction. There is no way for a mother monkey to communicate to her infant except by direct example what is a favorable spot for finding food, where to flee from a predator, or how to interact with other monkeys. She does not even communicate a message such as "Do this" or "Copy me." Description of the social and physical environment is beyond the scope of the communication system. We have emphasized that learning in a social context is a major part of the adaptation of the social primates because it encourages the development of a body of traditional, locally useful behavior patterns which may help a group to more effectively exploit its home range. It is all the more remarkable that this learning, which is so important to the development and survival of the individual as it matures, is not heavily supported by the communication system.

Aside from the example of predator alarms, social, nonhuman primates have little ability to communicate about their environment. Their way of life is such that this type of communication is very restricted, whereas exactly the opposite is true of *Homo sapiens* and human language. For monkeys and apes events inside the social group are of great importance, and their communication systems are well developed in their capacity to express motivation of individuals and to facilitate social relations. Without this ability to express emotion, monkeys and apes would not be able to engage in the subtle and complex social interactions that are a major feature of their adaptation.

HUMAN DISPLAYS

The more that is known about the communication systems of nonhuman primates, the more obvious it is that these systems have little relationship with human language but much with the ways our species expresses emotion through gesture, facial expression, and tone of voice. There is no evidence to suggest that human displays expressing emotional states, such as laughing, smiling, and crying, are any more or less complex than the displays of monkeys and apes, or that they differ significantly in form or function. For example, human weeping is a display which expresses distress and dependency and which often evokes solicitous responses from other human beings. Like the displays of other animals, it is based on a natural behavioral phenomenon which has undergone special evolutionary changes to increase its communicatory effectiveness. In this case there is an increase in lachrymal or tear fluid, a common mammalian response to protect the surface of the eye when the nervous system is under stressful stimulation. This protective response has been exaggerated in humans so that the fluid wells up and streams down the face in the form of tears. The facial expressions associated with weeping also heighten the effect of the display. This is an evolutionary development comparable to the whitening of the eyelids of many monkey species which adds to the effect of an "eyelid threat." In contrast to displays such as weeping or smiling, human language, a highly specialized aspect of the total human system of communication, has no obvious counterpart in the communication systems of our closest relatives, the Old World monkeys and apes.

There are many other human displays which are just as important to smoothly running social relations as language itself. Smiling, laughing, crying, weeping, screaming in fear, and roaring in rage are major patterns which are common to all human beings (Eibl-Eibesfeldt 1970). Even many relatively minor gestures and patterns such as the upraised palm in begging, the raised clenched fist of threat, and patting someone on the back to comfort seem to be nearly universal. We know that many of these patterns do not depend on learning for their expression. Take for example, smiling—babies born blind still begin to smile at the same time and in the same way as do normal infants. The smile is such a basic and vital part of human social interaction that its first appearance is not left up to chance learning experiences. If you doubt its importance, try to spend a day or so doing your normal routine, but do not smile at anyone. Soon people will begin to ask you what is the matter. If you refuse to smile when an acquaintance or coworker catches your eye, you are essentially refusing to recognize a friendly relationship. It will not be long before the relationship itself becomes distinctly strained.

Human displays are so important in communication that their meanings may even take precedent over language if there is a conflict. It is not possible to say "I love you" with narrowed eyes and clenched teeth and have anyone believe it. The facial expression has greater emotional force than "mere" words, and it is not surprising that we still lean heavily on these displays, or "body language" as they are now popularly called, when we are tying to comprehend other human beings' feelings and emotional attitudes.

HUMAN LANGUAGE, THE CEREBRAL CORTEX, AND THE LIMBIC SYSTEM

The most obvious difference between the human brain and those of the other primates is that man's brain is so much larger (Table 4–1). Whereas chimpanzees and gorillas have an average cranial capacity of 395 cubic centimeters and 500 cubic centimeters respectively, the average human cranial capacity is about 1300 cubic centimeters (Tobias 1963). In other words, the human brain is at least twice as large as that of his ape relatives. To interpret this size difference in terms of the behavior of the animals in question is quite problematic. There is a difference of up to 1000 cubic centimeters in the brain sizes of normal people, with no known behavioral consequences. Lenneberg (1967) emphasizes that brain size is no better a criterion for language function than it is for general intelligence. Certain kinds of idiots and small-brained dwarfs, both of whom have human brains about the size of a chimpanzee, are capable of mastering basic grammar and vocabulary, although their general mental abilities are severely limited. Such cases suggest that brain organization, not size, is the critical factor in determining language ability.

The mammalian brain consists of the brain stem, the cerebellum, and the cerebral cortex. The brain stem provides central control over all the basic functions necessary for life-respiration, digestion, and so on. The cerebellum assists in balance, locomotion, and fine motor control. The cerebral cortex receives and stores the body's sensory input, originates voluntary motor actions, and carries out all sorts of integrative functions. It is divided into two halves, or hemispheres, each of which transmits impulses along nerve fibers from opposite sides of the body. Three fiber bundles interrelate the two hemispheres.

In comparing the organization of the human brain with that of the other primates, certain trends can be recognized. For one thing, the surface area of the cerebral cortex is larger, due to folding during the course of evolution. This increased surface area permitted more cortical cells to develop, while the average thickness of the cortex remained relatively constant in all primates. This increase in cortical cells is not as great as it could be. As the surface area of the primate cerebral cortex increased, the nerve cell density decreased, and at the same time the branched ends that connect nerve cells increased. These features promoted more direct pathways and more differentiated areas of the cortex.

There is one crucial way in which the cerebral cortex of humans differs from those of other animals. In most animals the two halves of the brain are essentially symmetrical and represent mirror images of each other. If the left half of the brain of a dog were removed, there would be virtually no observable changes whatsoever in the animal's behavior. The same type of operation in an adult human would be disastrous—he would probably be left without the ability to use language for the

TABLE 4–1   RANGES AND MEANS OF CRANIAL CAPACITIES OF MODERN APES AND FOSSIL AND MODERN HUMANS (ADAPTED FROM TOBIAS 1963, 1971.)

Species	Size of Sample	Cranial Cubic Range	Capacity in Centimeters, Rounded Mean
Gibbon	86	87–130	90
Siamang	40	100–152	125
Chimpanzee	144	320–480	395
Orangutan	260	295–475	410
Gorilla	533	340–752	500
Australopithecines	6	435–540	494
Advanced Australopithecines	3	633–684	656
Homo erectus	11	750–1225	925
Homo sapiens (Modern humans)	1000s	±1000–2000	±1350

rest of his life. Studies of brain wounds, epileptic lesions, and surgical lesions show that a large part of the left cerebral cortex is involved with one or another aspect of language (Fig. 4–2). For example, Broca's area is a part of the supplementary motor cortex and it is concerned with the motor actions involved in producing speech. Wernicke's area is a part of the auditory association area and damage here results in a loss of the ability to understand spoken words. Sounds of words can be heard but cannot be understood. The angular gyrus or posterior language area involves the formation of associations between words and their referents, for ex-

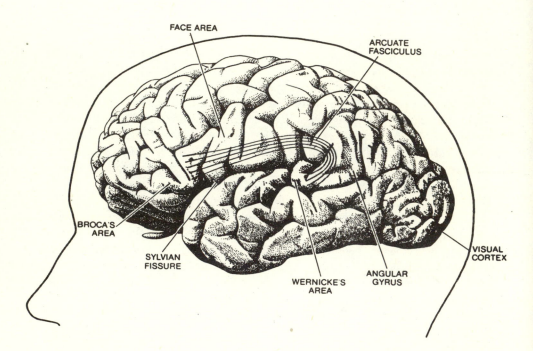

Figure 4–2. Primary language areas of the human brain. These areas are thought to be located in the left hemisphere, because only rarely does damage to the right hemisphere cause language disorders. Broca's area, which is adjacent to the region of the motor cortex that controls the movement of the muscles of the lips, the jaw, the tongue, the soft palate and the vocal cords, apparently incorporates programs for the coordination of these muscles in speech. Damage to Broca's area results in slow and labored speech, but comprehension of language remains intact. Wernicke's area lies between Heschel's gyrus, which is the primary receiver of auditory stimuli, and the angular gyrus, which acts as a way station between the auditory and the visual regions. When Wernicke's area is damaged, speech is fluent but has little content and comprehension is usually lost. Wernicke and Broca areas are joined by a nerve bundle called the arcuate fasciculus. When it is damaged, speech is fluent but abnormal, and patient can comprehend words but cannot repeat them.

ample, the connection between a visual or mental image of a dog and the actual sound of the word "dog." None of these areas have a functional counterpart in the right hemisphere. This assignment of language to the left hemisphere may be a solution of the problem of providing enough brain area for language without sacrificing other functions; language was assigned almost exclusively to one hemisphere while the other hemisphere acquired some nonverbal functions which are divided equally between both sides in other animals. Lateralization in humans, then, represents the complex basis of brain reorganization necessary to develop language—a learned, object-oriented system of communication existing alongside of the old, emotion-based set of signals common to animals and human beings.

The ability to name objects by using structures in the cerebral cortex provided humans with a way to refer to their environment with each other. It is important to stress that this ability would carry with it a measure of independence of the signal from the limbic system and the direct, unconscious expression of individual emotion. A freeing of vocalizations from the emotions and a lowering of the threshold for vocalizing are major prerequisites for human language. The rare examples of nonhuman primates communicating some information about the environment in their high-intensity alarm calls are not relevant to the evolution of language. The similarity between these abilities is only a slight one at the level of behavior, and the underlying mechanisms are entirely different. It is apparent that the ability of the vervet monkey to refer to the environment, a design feature which is perhaps behaviorally similar to human naming but which is in fact based on quite different underlying neurological mechanisms, cannot be suggested as representing a possible step toward language. The behavior of naming, as it is used here, is uniquely developed in humans because the anatomical basis of the ability is also uniquely developed in the human cerebral cortex.

It is clear then that human beings can and do make two different kinds of vocal responses to a stimulus. There is a great difference between crying out or screaming in response to seeing a lion and saying, "There is a lion." The first kind of vocal response is an expression of the emotional state of the individual at that particular moment. It is based on a primitive brain mechanism common to all mammals in which the image, sound, or smell of a lion stimulates a part of the limbic system. The act of vocalization represents the fact that the limbic system has just been stimulated. In contrast, when a human says, "There is a lion," the fact of the vocalization does not necessarily indicate any significant degree of limbic stimulation. Instead, a uniquely human neural mechanism involving the cerebral cortex is used. Often both systems are stimulated at the same time and constriction of the larynx will change the quality of the sound so much that an emotional state will also be communicated as the person is speaking. However, if stimulation to the limbic system is extremely strong, it often precludes the use of the cortical connections for language and a person may become speechless with fear or excitement. The difference, then, between a system of emotional signals and human language is a qualitative one. Not only are they different in form and function, but they also differ in respect to the neurological mechanisms upon which they are based.

It will be noticed that no mention has been made of the anatomical mechanisms used in making sounds. This deliberate omission comes from the conviction that no evolution in the mouth, tongue, or larynx was necessary to the origin of human language. Nonhuman primates make a variety of sounds. The initial question in the origin of language is how our ancestors started using some of these sounds as names; the particular sounds are not important. After the learning of names had brought in new selection pressures, there may have been evolutionary changes of the sort described by Lieberman (1968), in sound-producing structures and in the nervous control of lips, tongue, and laryngeal muscles. The argument is essentially the same as that on the relation of tools to the brain and hand (Washburn 1960; Lancaster 1968a). The adaptive success of early tool use changed selection pressures on the human organism. Behavioral changes probably preceded many anatomical or structural changes, but once their adaptive value was established, then the area of cortex devoted to hand skills greatly increased and the structure of the hand itself, especially in the length and muscularity of the thumb, evolved for more efficient opposition to the fingers. In such an evolutionary sequence a mutually stimulating, feedback relationship exists between structure and behavior which continues over time. As the system becomes more and more effective, more and more refinements of the supporting anatomical structure can occur. In the same way, human language might easily have evolved based on a range of sounds different from those used today. The rudiments of naming and grammar could not evolve without fundamental evolutionary changes in the structure of the brain. This evolution expresses itself today in the relative ease with which the human child learns naming, grammar, and the whole complex of language, and in the difficulty with which a monkey or ape learns any aspect of what might conceivably be called human speech. The ape makes many sounds. It sees much as we do. Its facial expressions are rich in number and meaning. But the structural basis in the brain for combining these abilities into meaningful, spoken language is lacking.

TEACHING "LANGUAGE" TO CHIMPANZEES

Within the past few years a number of studies on the chimpanzee, both in the natural habitat and in the laboratory, have brought to light their amazing mental abilities. Van Lawick-Goodall has described an astonishing variety of tool-using behavior patterns displayed by chimpanzees in the wild. She has seen the use of stick and grass probes for catching termites and ants, leaf sponges to collect drinking water, leaves for wiping dirt from the body, and objects like sticks or stones thrown in aggressive display (van Lawick-Goodall 1970). As you know, chimpanzees not only use tools but they also make the tools themselves. Van Lawick-Goodall has seen chimpanzees make a termite probe by selecting a long thin stick, stripping off any side twigs or leaves, and then breaking it to the proper length. Crude as this tool-using and tool-making behavior may seem by human standards, it is nevertheless unrivaled in both variety and complexity by any other animal species aside from humans.

Further evidence on the mental abilities of chimpanzees has come from two pioneering laboratory studies by Gardner and Gardner (1971), and Premack (1971). These researchers have been working independently but they began with similar premises and have arrived at somewhat similar results. Both studies began with the idea that any attempt to teach human language to a nonhuman primate should utilize a channel which a primate can use easily. The Gardners noted the freedom of movement chimpanzees have in their hands and decided to use American sign language. Premack devised a language system which made use of plastic symbols on a magnetized board. Both of these workers found that after several years of training the two chimpanzees had built up "vocabularies" of 85 symbols in one case and about 100 in the other. These two now-famous chimpanzees, Washoe and Sarah, can not only comprehend the meanings of these symbols but they can also produce their own messages, sometimes applying them to novel situations and in new combinations. Premack has been able to go so far as to teach Sarah to respond correctly to messages which make use of both a negative concept and a conditional clause. For example, Sarah, who prefers chocolate candy above all, responds correctly to plastic symbols arranged on a board to mean "Sarah take apple, then Mary no give Sarah chocolate." Premack did not teach Sarah such complex concepts as conditionality and negativity. As he has pointed out many times, language training is mainly a mapping of existing knowledge. Language gave names to distinctions which Sarah's ape mind was already capable of drawing. According to Premack, he could only teach language to Sarah when the sentences used in training perfectly encompassed distinctions in her prelanguage experience.

The ability of the chimpanzee to comprehend and use symbols must be understood in its proper context. It must be emphasized that there is no evidence that chimpanzees use anything like language to communicate with each other in the wild. Van Lawick-Goodall has been studying a group of wild chimpanzees for 13 years now in Tanzania. She has thousands of hours of observations on social behavior. Van Lawick-Goodall (1968) and Marler (1969) have studied the gestures, movements, and vocalizations used by chimpanzees to communicate with each other. There is no suggestion of anything resembling either words or sentences. Chimpanzees use signals expressive of emotional states just as do monkeys and other animals. No matter how complex the behavioral abilities that the Gardners and Premack have revealed in their laboratory studies, we know that they do not relate to the way in which chimpanzees communicate with one another. These abilities play some other role in the adaptive pattern of the chimpanzee at which for the moment we can only guess. It may well be that the selective advantage of these abilities somehow relates more closely to their tool-using behavior than to their communication system. For example, to make a stone chopping tool one must have a mental concept of the finished product before striking off the first flake from a lump of stone. Perhaps tool making and using was an important selective agent for the ability to form associations between an abstract concept and a tangible object, which in turn was a necessary precondition for the evolution of language itself.

The experiments of teaching language to Sarah and Washoe have brought to light some interesting similarities and contrasts between apes and humans. Although both chimpanzees now use language at a level comparable to that of a two and one-half to three year old child, there is a marked contrast between how a human child acquires language and how these two chimpanzees did. In fact, language is not taught to a human child. It is virtually impossible to keep a normal child from learning language unless he is raised in total isolation. Human children do not have to be taught syntax and grammar; they will unconsciously appreciate and apply all of the major types of sentence structure and grammatical rules of their native dialect before they ever go to school. Language is part of the biology of the human species just as is walking erect on two feet. Neither of these activities have to be taught and only the most extreme forms of deprivation or malformation can keep a human child from talking and walking.

Another important contrast between chimpanzee "language" and human language comes from the sensory modalities used. The Gardners and Premack used the visual mode because they knew that chimpanzees relied heavily on sight in sensing their own world—both physical and social. The Gardners had originally hoped that Washoe would acquire word gestures without special training simply by being with human beings who used American sign language to communicate with each other, and kept up a running commentary on the events of the day as people often do with small children. They found that this was extremely difficult because so many activities use the hands, the body, and visual concentration. In contrast human language, using the mouth and ears, interferes with almost no activity except for active eating or listening to something. Human language, founded on the vocal-auditory channel, was ideal for the evolving human species as it developed new behavior patterns, such as tool use and manufacture and hunting, at the same time that language probably evolved. Rather than interfering with each other, these new activities were complementary and mutually supportive. It must be emphasized that language is really a communication system and not a thought process. The mental processes of human and nonhuman primates share many features such as appreciation of symbols and order. The adaptive value of language in the course of human evolution lay in the fact that it was a communication system which encouraged the exchange of information about the environment and it used a vocal-auditory channel, leaving the eyes and hands free for tool using.

THE EVOLUTION OF ENVIRONMENTAL REFERENCE

The fossil record gives clear evidence of a tripling in size of the human brain during the Pleistocene. Comparative neuroanatomy suggests that this was especially true of the cerebral cortex. Although the fossil record itself gives little evidence, it is unlikely that this increase was a generalized one in total size. It is a reasonable assumption that much of the expansion was disproportionately in favor of special areas of cerebral cortex, areas which play a major role in determining behavior patterns peculiar to the human species. These areas may have been very small or not represented at all in the brains of our ape ancestors. Their expansion was most likely due to selection pressures caused by emerging behavior patterns which are now thought of as typically human. A good proportion of this increase was probably in the posterior speech area (see Figure 4-2) (Lancaster 1968b). Other changes at the cellular and biochemical levels undoubtedly paralleled those at the neuroanatomical level, but these leave no trace whatsoever in the fossil record. The expansion of this region gave humans the ability to form an almost unlimited number of associations between the sound patterns used to name and their referents in the environment. The ability to form such associations lies at the heart of human language and differentiates it from other primate communication systems. The communication systems of nonhuman primates rarely make any reference to the environment but rather they express the emotional state of individuals. In contrast, language provided human beings with a tool by which they could communicate information to others not only about their own emotional states but also about social relationships and the physical environment.

Human language is both universal and species specific. All known human languages as far back as we can trace them are organized on the same basic plan. Language is everywhere transmitted by the vocal-auditory channel, although additional secondary modes of transmission, such as writing, are now common. Language is hierarchically organized, with referential names forming its core. These names are made up of combinations of a small number of basic but meaningless sounds which differ from language to language. Referential names rarely occur in isolation; they are typically combined into sentences according to specific rules (grammar). Although each language has only a few basic sounds, these sounds are combined productively, to produce new names and especially new sentences.

An ability to make environmental reference, even on a very primitive level, would have had great adaptive value to early humans providing it was ecologically important for them to communicate information about the environment among members of the group. Even a very small vocabulary of object names, which might have been combined in simple utterances of only a word or two, could have been very effective. A few words suggesting major activities such as gathering or hunting and a few others indicating relative nearness in space and time, in combination with a small vocabulary of names, could have had important adaptive significance.

The ability would have greatly facilitated gathering and hunting with the use of tools. The way humans use tools is an activity just as uniquely human as is language. This is probably no coincidence. It is worth considering whether the one unique ability, tool use, might not be closely related to the other, language—and especially environmental reference. Tool use in itself tends to focus attention on objects in the environment so that a stone, stick, or reed that has no significance or meaning in the life of a monkey, or at least none that need be communicated to others, may have great importance to the maker of a primitive pebble tool or a woven collecting bag. Naming means that simple but highly significant communications can be made in which raw materials or tools can be referred to and animals and plants named.

The ability to make environmental reference can greatly alter the way in which a group can exploit its home range. If a social group can split up into subunits that forage separately during the day or for a few days and then join again, they can utilize the food resources of a large home range much more efficiently. The chances of locating rare, dispersed, and seasonal food sources is greatly increased. But such an ability presupposes a place to meet where food can be shared with those who were not successful in their gathering or hunting. A permanent home base would limit the potential size of the home range but, if a group had a way to designate a future meeting place, a home range of hundreds of square miles could be utilized effectively. Even on the rudimentary level of a simple place name, naming could have had great selective advantage to an early hominid form that was moving into a way of life based on tool use, gathering and hunting, and food sharing. It must be stressed, however, that tools, gathering and hunting, bipedalism, and naming formed a unified complex which over time led to the evolution of a new way of life. Selection was for the whole successful way of life, of which environmental reference and naming were an important part. But just as the behavior patterns of language, gathering and hunting, and tool using made a more complex way of life possible, at the same time all of these new complexities in human experience fed back onto the evolution of the brain. No single evolutionary advance had value without the rest of the pattern and no single element can be entirely understood alone. Rather, tools, language, and gathering and hunting must be valued as major parts of a single adaptive complex.

The ability for environmental reference both reflects and permits a relationship between the species and its environment different from that of other primates. The emergence of language was related to changes in the ecology of the evolving hominid species. According to this view, the evolution of the abilities to use tools and to name are closely linked. These abilities form an important and very early human specialization which, even at its earliest and most rudimentary level of development, could have had great adaptive significance.

The formation of associations between sensory images and names is basic to much of what are understood to be human mental abilities such as abstraction, symbolization, and classification. Animals certainly demonstrate some ability to abstract and classify. For example, many species of birds and mammals are able to form limited numerical distinctions (Thorpe 1963) up to perhaps seven or eight without the aid of names. Human beings also seem to be able to appreciate numerical differences without counting up to approximately the same number. However, humans can also give numerical names to quantitative differences, and this skill gives us an unlimited ability to count. Numerical abstractions, which are one form of classification, are unlimited in humans because they can give names to them.

It should be emphasized that the origin of language is a separate problem from understanding the language of human beings as it is today. We are not attempting to explain all of the complexities of human language, its multiple forms, and its varying functions. Rather we are trying to understand what language could have been like before it was the highly evolved system of communication of today. We are emphasizing what we consider to be features of language which evolved very early in human history and which provided the base upon which later refinements have been built. Human speech represents a major complex of behavioral, neurological, and anatomical elements. This complex neither sprang from whole cloth nor did its parts necessarily evolve at a smooth and even rate from a primitive to a highly advanced level. Parts of the complex may have been highly evolved long before other parts began to differ at all from their primitive state. This chapter, then, does not try to explain or discuss all of human language but does attempt to interpret and evaluate the facts that are relevant to understanding its emergence.

TOPICS FOR DISCUSSION OR PROJECTS

1. Make observations in the classroom, dormitory, or cafeteria on selected aspects of human displays such as those used in greeting, parting, or in the expression of affection or aggression.

2. Think of ways in which language with a referential function changes the relationships between individuals or between human beings and their environment.

The Human Adaptive Pattern

ADAPTIVE COMPLEXES IN EVOLUTION

It should be evident by now that major evolutionary changes usually involve whole complexes of adaptations at many different levels—behavioral, anatomical, physiological, and social. Adaptive changes do not occur in isolation but are bound to affect other parts of the system. Earlier in the book, we touched upon a number of adaptive complexes significant to the evolution of the primates. For example, the evolution of the grasping hand involved major changes in locomotion as well as feeding behavior. The evolution of language involved a major biological complex including changes in social behavior, ecological adaptation, and the neuro-anatomy of the brain. The evolution of tool using and tool making again called for major alterations in locomotion (bipedalism), in the hand (the well-muscled thumb used in opposition to the fingers) and in the brain as well (control of skillful hand movements).

Often various elements in an adaptive complex develop in a feedback relationship with each other as they evolve together through time. It is easy to understand this relationship in the case of the evolution of bipedalism. Bipedalism in itself is not a very efficient way to move about. Humans are clearly at a disadvantage both in terms of speed and balance. If you doubt this, try racing over rough terrain with your dog. Furthermore, the anatomical changes necessary for efficient bipedalism are not simple and probably included some rather awkward intermediate stages. Nevertheless, bipedalism did evolve perhaps partly because of its feedback relationship with tool using. In the beginning, tool using was probably quite casual, as it is for chimpanzees. Tools were probably more or less made on the spot and discarded when the task of the moment was over. But, when a particularly useful or scarce bit of sharp stone or a well-balanced club was found, there was clearly value in carrying it. The apes are quite capable of a kind of bent-kneed, bipedal shuffling walk, not the efficient forward stride of true humans, but still bipedal. Tool using and the carrying of tools or food gave adaptive value to this minor behavioral trait in the apes. Gradually through time the shuffling biped who used tools casually became the efficient biped who was fully committed to a tool using way of life.

As was stated in Chapter 1, the earliest clear evidence of the human pattern comes from the Australopithecines. The fossil and archeological record of these creatures gives evidence of a number of the major features of the human pattern, such as almost fully evolved bipedalism, the making of stone tools, the killing of game, and home bases or campsites. The time span of these early humans was from over 5 million until about 1 million years ago. All these archeological features are not represented at the earliest dates, but elements of the pattern are clearly present, suggesting that it is only a matter of time and luck in recovering more fossils before the entire period is filled in. At the moment, the time period between 5 and 10 million years ago is nearly blank for both human and ape fossils. However, fossil-bearing geological formations have been found from this period and it is probable that in time the human lineage will be traced back into this period. Eventually it will come to a point where fossils will be found that cannot be clearly classified as either an ape or an Australopithecine but something in between.

There are many features of the human adaptive pattern, particularly those involving changes in social life, that do not leave clear evidence in the archeological record or can only be guessed at from slim evidence. Nevertheless, these changes in social life were just as important a part of the emerging human adaptive pattern as were stone tools. We can use our knowledge of the behavior of nonhuman primates as well as the lives of people living today in a gathering-hunting economy to help reconstruct what these changes might have been.

Let us take what we know about nonhuman primates and try to reconstruct the most probable behavioral and social attributes of our ape ancestor. We may be wrong on some of these guesses, but at least they are based on the best possible evidence available today. First of all, we would probably find some important differences in the behavior of males and females. Primate field data draw a picture of an adventurous, wandering male able to form bonds with neighbors and relative strangers, live comfortably in unfamiliar home ranges, and vigorously defend new territories. These males would have a strong tendency to take up defense of females and young against outside threat but they would take no economic responsibility for the well-being of females and young.

Females, in contrast, would be strongly bounded to their genealogies, so strongly perhaps that, if an old dominant female left the group, her genealogy might follow her. However, a female would be unlikely to leave the home range of her birth, and her interests would be strongly focused on her offspring and close relatives. A female's genealogy might contain three or four generations and consist of both males and females. High ranking and often larger genealogies might form the social core of the group and the mature generation of "elders" might be composed of a strongly bonded group of brothers and sisters. Both males and females would be equally concerned with questions of dominance and social rank. However, for a number of reasons, some of which are still unclear, males were more likely to achieve a rank which differed from that of their genealogy and so the relative rankings among adult males would remain unstable through time. Females, on the other hand, would have maintained stable rank orders which endured for long periods, and major changes may have only occurred when high ranking genealogies were decimated by death, disease, or by a skew in the sex ratio of births in favor of males.

Because of the stability of their social relations and their attachment to their home ranges, adult females may have had much greater knowledge about the

Figure 5–1. Mike, the dominant male of the Gombe group, begs Old Leakey for a piece of meat. Although Mike has both the physical and social power to take the meat away from Leakey, he does not. Chimpanzees rarely fight over meat, rather they share. The empty palm gesture for begging is common to both chimpanzees and humans.

resources of their environment than did the males, and they probably played major roles in group decisions. Individuals were largely responsible for feeding themselves, even infants once they were weaned. The diet consisted mainly of vegetables and fruit, but there was a definite appetite for animal protein, usually in the form of insects, small reptiles, young birds, eggs, and other easily collected items. On occasion some of the adult males hunted larger game and often they would cooperate in hunting down prey. Sometimes this meat would be informally shared among group members nearby at the time (Fig. 5–1). Both males and females used tools casually, much like wild chimpanzees do today. Tools were not skillfully made, but they were used in a wide variety of circumstances.

If the above is a reasonable reconstruction of the behavior of our ape ancestor in respect to social organization, then it is clear that many of the essential ingredients of the human adaptive pattern were foreshadowed in the behavior of nonhuman primates. The next question is what happened to transform this creature into a human being. The human adaptive pattern rests on a fundamental shift in economic and social relations among members of a social group. This shift was distinctly human even though it was based on a primate heritage and foreshadowed in the behavior of our ape ancestors.

THE MAJOR FEATURES OF THE HUMAN ADAPTIVE PATTERN

The Division of Labor

Many of the elements of the human adaptive pattern have now been found to occur in the behaviors of other primates. In some cases these patterns are major adaptations even in nonhuman primates, such as the establishment of the matrifocal lineage or family in many monkey and ape societies. Other crucial elements of the human pattern such as bipedalism, tool using and tool making, food sharing, and cooperative hunting of meat do occur in nonhuman primates but only as relatively minor behavior patterns and not as key adaptations. These elements came together in early humans coupled with a fundamental change in social and economic relations within the group. This new pattern of behavior opened up an unexploited niche for human beings—one that must have been relatively protected from competition so that even the small-brained Australopithecines were very successful living in it.

More than anything else the new pattern must have rested on a unique way of exploiting food resources: gathering and hunting based on a division of labor between adult males and adult females with offspring. As was explained earlier, in nonhuman primates each animal is a separate subsistence unit. Infants may be dependent on their mothers and their groups for protection from other monkeys and the dangers of the environment but, once they are weaned, they must feed themselves. In contrast, weaned human young are dependent on adults not only for protection and security but also for food for many years. Because of this long-term dependence on adults for nourishment, the roles of both male and female humans have greatly expanded and much of their time during each day is spent in activities which provide food for the young.

In nonhuman primates finding food means foraging for fruits, vegetables, grasses, and sometimes insects, eggs, and other sources of animal protein. Humans added to, but did not lose, that gathering pattern when they became cooperative hunters. Because of the long-term dependence of children, a division of labor evolved in which the adventurous, wandering male became the hunter and the female developed the less mobile role of gatherer and mother. The home ranges of group-hunting mammals are always very large compared to foragers like monkeys (Schaller and Lowther 1969) and hunting can be a dangerous activity demanding skill and concentration. As Brown (1970) has so ably pointed out, there are certain activities which cannot be done by an individual burdened with immature offspring. Child care responsibilities are only compatible with activities which do not require long trips far from home; with tasks which do not require rapt concentration; and with work which is not dangerous, can be performed in spite of interruptions, and is easily resumed once interrupted.

In time past mother's milk was the essential food for infants for the first year or so of life; normal growth and development depended on it and there was no substitute (even the artificial milks of today are inadequate). Since the care of infants had to fall to females anyway because only they could feed them, it is obvious that care of all immatures would come to the women in a gathering-hunting division of labor. Furthermore, hunters can not be burdened with children. It is impossible to go on a three-day hunting trip which may cover a hundred miles or more and carry an infant with a three-year-old following along behind. Clearly, with the long years that it takes for the human child to develop and learn adult roles and skills, there was no other way once gathering and hunting had developed as a major adaptive stance for the division to have evolved except between males and females. There is no need to posit special "killer" or "maternal" instincts in males and females to explain the assignment of these roles.

There are two very important benefits from the division of labor. The first is probably the key to the human species' early successes long before the evolution of large brains and culture. The division of labor provided a flexible system of joint dependence on plant and animal foods (Isaac 1968, 1971). It created an effective pattern very different from that of group-living carnivores. A stable and balanced diet was certainly much more likely. Hunting is not necessarily a successful enterprise on a day-to-day basis and data from modern hunters and gatherers indicate that meat usually constitutes less than 50 percent of the total diet (Lee and DeVore 1968). Carnivore populations are subject to bad years, but the human gathering-hunting pattern has tremendous flexibility. It can adjust to daily, seasonal, or cyclical variations in food supplies as well as geographical and ecological variants. This system permitted the human species to cover most of the earth without speciating, that is, without having to make major changes in anatomy and behavior in order to maintain an adequate food supply in different ecological zones.

The second advantage of the division of labor was that it marked the first step in specializations in food-getting activities. The tools and the way they were used differed between males and females—the beginning of the evolution of more skilled performances in a variety of tool-making tasks and food quests. For example, the effective use of projectiles such as spears, darts, and arrows takes years of practice. Much of the play of young humans involves the development of skills they will need as adults. If only half the adult repertoire needs to be learned by any one individual, then chances are that the learning of such skills will be more efficient.

It should be remembered that the division of labor involves some very complex behavior patterns to make it effective. The first of these is a psychology of sharing on a regular basis, not just after the finder of food has been satiated. As was mentioned earlier, in spite of their intelligence monkeys and apes do not think about the nutritional needs of others. Apes sometimes respond to begging gestures (see Fig. 5–1), but it would never occur to an ape mother to put something aside for her youngster to be sure it has enough to eat that day. For food sharing and the division of labor to really work, food has to be shared especially when times are bad and some foragers come back with nothing and others with just a little. Washburn and Moore (1974) suggest that food sharing may have placed a new adaptive force on the human species—one that selected against groups with overly aggressive and dominant individuals who could not control their own emotions and need to dominate others.

Even with a psychology for sharing there still are some critical logistical problems that have to be solved before life or death can depend on it. When individuals of a group separate during the day, some to gather and others to hunt, they must have some way to meet again later. This is not much of a problem if the home range is small and the number of sleeping areas limited, but hunters have home ranges containing hundreds of square miles. Critical, then, in food sharing is the designation of a home base, perhaps just a camp site used for a month before the whole group moves on. The archeological record shows such sites for the Australopithecines, areas that were occupied for more than a single night where food refuse, broken tools, and waste from tool manufacture accumulated. The evolution of language must have helped humans a great deal in utilizing various bases in their home range because they could talk about them—refer to places and activities that were not present at the moment but somewhere else and in the future.

A final logistical problem in food sharing and the division of labor is how to carry the things to be shared. As the van Lawick-Goodall movies show, carrying five or ten bananas is no easy task. Chimpanzees will load up with bananas, sticking them under the arms and chin, between the teeth, and between a flexed leg and groin. Carrying, for the chimpanzee, is a major task taking up most of his concentration and effort. The adaptive advantage of such simple inventions as a skin to carry water or a fiber mesh bag to carry nuts or fruits must have been just as great as the more dramatic development of tools like knives or spears. In fact, the traditional tools of gathering women (the digging stick and the carrying bag) are never found in the archeological record because they do not preserve well. This is one of the reasons why Stone Age archeology has traditionally focused on the activities of males as the major factors for subsistence survival—compared to the females, the activities of males more often left records in bits of broken bone or stone.

The Human Family

The uniqueness of the human adaptive pattern does not rest on the division of labor alone. Human males do not relate to females and young simply as members of an age–sex class. The human pattern is one in which specific males relate to specific females and their young, taking on special responsibility for their welfare. This relationship and its obligations is the male role of "husband-father," a role with no true counterpart among the nonhuman primates (Fig. 5–2). There is no way to prove when this role first appeared but there are good reasons to think that it was closely linked to the establishment of the division of labor. A division of labor without a husband-father can have some very serious side effects, especially in a small group. Take for example a small group of 25 individuals, over half of whom will be immature. Chance alone will cause major skews in the sex ratio among adults. How can a division of labor work in a group of ten adults if two are female and eight are male, or vice versa? This skew could be a disaster during a drought when the only available food was from hunting. With two males having to provide enough food for eight women and all their children, even if hard times only lasted for a month, the group could lose members through starvation or disease. This kind of skew will happen through chance variation in births or from misfortune and disease. There is one ideal behavioral mechanism which will guarantee that the division of labor occurs between reasonably balanced numbers of each sex—the establishment of the role of husband-father. This assures each female that there is a male hunter in the group to balance the needs of herself and her children. It also insures that extra, unmated hunters will look in neighboring groups for females so that a whole region becomes a pool for finding mates instead of just the local group. Today by far the most common form of family in small gatherer-hunter groups is monogamous (one man–one woman) simply because most men can only really hunt for one childbearing woman at a time. However, a few men do take on extra wives if they are exceptional hunters or if the second wife is an older widow, no longer bearing children. Generally there is a relative shortage of men because in

Figure 5–2. A family picnic without father: Flo's family. The old female Flo lounges with three of her living offspring: Fifi, a young adult female; Flint, a juvenile male; and Flame, a six months old female. Fifi is grooming Flo's arm in an attempt to handle or touch the infant.

most human groups males have a slightly higher death rate than do females right from the moment of conception.

The human family represents the bonding of an adult male to a matrifocal unit for economic reasons. However, the bonding itself is not a rational, economic process but rather an emotional one. Important adaptive relationships are not left to simple experience and rational evaluation alone to keep them going, but rather they are strongly supported by the emotions. So in humans we find major changes in the relations between the sexes which help to maintain this vital bond. The most basic of these changes is in human sexual activity. Sexual activity acts as a social glue, so to speak, not between males and females as a class, but between mates. For this to happen, a major change in the reproductive patterns of the species had to occur—the suppression of estrus.

All female mammals have special changes in the body and in behavior around the time of ovulation which make them seek mates and make them attractive to males. Estrus in a couple of mature females will bring a chimpanzee group to an orgy of copulations lasting several days in which all the males of the group copulate several times with each estrous female. After that there may be no sexual activity for weeks or months until some other females come into estrus. What would happen to the division of labor if human females came into estrus? What if times are bad and food scarce; who is going to go hunting if there is an irresistible female in camp? But more important, what happens to the special bond between mates under these circumstances? The human adaptation has been the suppression of estrus and continuous sexual receptivity under conscious control in both partners. Even so, vestiges of estrous behavior still exist in the fact that most human females seem to be more sexually oriented around the time of ovulation, though they may not be conscious of its occurrence. Parallel to this have been changes in the factors that elicit sexual excitement in the human male. A human female does not have to be in estrus to be sexually attractive—instead she has evolved a series of special "sign stimuli" which mark her as sexually exciting all of the time. These special features, such as fat pads around the nipples and on the hips and buttocks, and relatively hairless skin except for the head and pubic area, are unique to the human female and have evolved to take the place of estrus of nonhuman primate females.

It is interesting to note that humans are generally not a very sexually dimorphic species. If we take a measure of sexual dimorphism, such as in body weight, we find that man represents a minimal difference of about 10 percent between males and females (see Table 2–1). Height and body weight dimorphisms usually relate to major differences between the sexes in behavioral potentials for aggression, as in the baboon and gorilla. The fossil record suggests that even as far back as the Australopithecines sexual differences in body size and weight were minimal in the human line. Differences in muscularity between men and women is greatly exaggerated in modern society because of the relative inactivity of females. As Brown (1970) noted, after reviewing cross culturally the assignment of tasks to women, differences in strength between the sexes were never taken into account, only compatibility with child care. Aside from the minimal 10 percent development of sexual differences in body size, sexual dimorphism in humans relates to sign stimuli like beards or breasts.

There was another adaptive advantage from the family as a subunit in the group or band similar to that of the harem system of the hamadryas baboons. The family represents a minimal-sized foraging unit with a balanced membership. It is common among modern gatherer-hunters for family groups to separate and forage independently when game is scattered, and to gather together when communal hunting is possible. For example, among the Bushmen in Southern Africa, both humans and game collect around water holes during the dry season. This is a time for hunting and for families to join together into bands. Relatively little time and effort need be put into the food quest, and social activities such as curing ceremonies, dances, and feasts predominate. Once the rains come, the band breaks down again into its smaller component units, families, which scatter into the bush where men will hunt and women will gather in small groups. Similar seasonal changes involving congregation into bands and then dispersal in family units have been reported for many other gatherer-hunter groups (Lee and DeVore 1968).

The adaptive advantage, then, of the human family may have first come from the fact that it helped to maintain relatively balanced groups in order that the division of labor could function effectively. In fact, it is difficult to see how a division of labor could have worked if there was no behavioral mechanism to keep a balanced sex ratio in these small groups. A division of labor would be a distinct disadvantage to most groups if the sex ratio were unbalanced, because it would keep many adults from fully productive activity at certain times of year even though they would still have to be fed. Two women gathering for eight men is just as inefficient as two men hunting for eight women—there are bound to be times when the group's food supply is inadequate. In certain environments gathering can take just as much experience and skill as hunting; once a division of labor has been made it cannot be simply undone overnight.

There is no real archeological evidence to prove that either a division of labor or a human family system existed in the Australopithecines. We do know for certain that the Australopithecines walked erect, made crude stone tools, had home bases, and hunted meat for food. We can only guess that the other elements of the adaptive complex (the division of labor and the family) were there as well. The division of labor and the family represented major economic and social changes in the common primate pattern. They may explain how the Australopithecines, in spite of their ape-size brains, were able to compete successfully with their brother apes and establish themselves in a new adaptive niche.

SMALL-SCALE SOCIETIES AND THE MODERN CONDITION

Millions of years have passed since the special human pattern of social and economic relationships first evolved. The Australopithecine stage of human evolution, marked by small-brained forms and minimal development of culture, ended around a million years ago. The Australopithecines were replaced by a much more advanced form of human being, *Homo erectus* who was himself replaced by modern humans, *Homo sapiens*. These later forms of humans differed from the Australopithecines in the size of their brains. In less than a million years the human brain tripled in size in conjunction with a tremendous proliferation and elaboration of culture. Nevertheless, humans were still gatherers and hunters. The earliest known evidence for the domestication of plants comes very late in human history, around 12,000 years ago in Southeast Asia. Even the domestication of plants and animals did not radically change human experience in terms of many aspects of social life. Most humans continued to live in small groups, mainly interacting with well-known individuals, many of whom were kin.

Major changes in human social and emotional experience came with the industrial revolution in the nineteenth century and many parts of the world are only feeling its effects today. It is true that in parts of Africa, Asia, Europe, and the New World cities have existed for thousands of years, so that some of the sweeping changes of the twentieth century were foreshadowed for certain limited populations. However, many of the discontinuities in social and emotional experience that we are going to discuss are not due to urban life alone but to other factors only felt recently. Furthermore, in the context of evolutionary change a few thousand years is very little time to alter adaptations which have been constant and universal for millions of years. Remember that 99 percent of human history was spent in a gathering-hunting economy with relatively stable and isolated populations.

In Chapter 4 we talked about the adaptive nature of the emotions—how they make the individual want to do what he has to do in order to survive, reproduce, and care for offspring. For example, in many societies of monkeys and apes social grooming is a major behavior pattern taking up many hours of the day. The grooming serves to keep coats clean and free of parasites, but far more importantly it makes individuals stay together, have physical contact, and interact positively (Fig. 5–3). However, none of these reasons explain why the individual actors groom each other. We have to understand their behavior as motivated by the pleasure or satisfaction it gives to both the groomer and the groomed. In other words, just as with human sexual behavior the motivation of the actors involved is very different from the adaptive value of the behavior itself. The emotions provide the short-term reward for the behavior, and evolutionary success maintains the behavior pattern in the repertoire of the species. The social emotions have evolved to fit individuals into the kinds of social systems which have proved successful for the species. So, like other kinds of adaptive traits, the emotions represent successes of time past and may not necessarily fit the circumstances of time present. For this reason it is worth considering some of the major contrasts between social life in a small-scale society and human experience in the modern urban setting.

One of the most striking and far-reaching contrasts has been in the increase in life expectancy. Unless there are some radical changes in human behavior, the world population is expected to double by the year 2000. There are two reasons why we are experiencing a population explosion: first because more children survive their first three years of life and secondly because people are living longer. The greater survival rate of children ultimately will have the greatest impact because it affects the reproductive potential of future generations. In most nonhuman primates adult females give birth every one to four years (depending on the species) from sexual maturity until death. Although a female monkey or ape can easily produce 10 to 15 offspring during a lifetime, most of these will die before sexual maturity. In fact, a number of field studies have reported 50 percent infant mortality during the first

Figure 5–3. The pleasure of grooming. Social grooming is not a service related to dominance but a social act pleasing to the groomer and the groomed. In this case one of the highest-ranking females in a group of vervet monkeys grooms her adult daughter. Her daughter, also highly ranked, grooms the least dominant of the 18 adult females in the group.

year of life. Among humans, too, the death rate of infants has been high in time past due to factors such as infectious disease and parasites. The high death rate of infants, combined with a birth spacing of two or three years due to low fertility during lactation, meant that in time past most human populations were relatively stable or only expanding slowly. Today twentieth century advances in public health have practically erased infant mortality in modern countries, and even in underdeveloped countries the drop in infant mortality has been startling. In spite of this, human sexual activity continues at the high rate to which it is adapted because its evolutionary value has been as a bonding mechanism between partners cooperating in the raising of children. If sex among humans were for procreation alone, its frequency would be minimal. A woman might only have intercourse for a few days once every few years as do females in some monkey and ape societies. Fortunately, in this one case modern science and technology have provided a solution to a problem which they created. Effective and inexpensive birth control techniques have separated the two functions of sexual activity—social bonding and procreation—so that it is now possible to have the one without the other. The effectiveness of birth control, of course, rests on the motivation of individuals to use it. However, the 1970s have witnessed such radical cuts in birth rates in parts of Western Europe, the United States, and China that, if continued, these countries will have no more growth in population.

The second reason why populations have been expanding is because people are living longer. This increase in life expectancy of adults has some interesting effects on social relationships. For example, today when a young couple aged 20 form a pair and begin a family, they can expect to spend as much as 50 to 70 years together, with perhaps two-thirds of their marriage spent in the nonreproductive phase. The traditional division of labor worked well in most situations right up into the twentieth century. But today, with long life expectancy combined with birth control, a woman may spend only a small fraction of her lifetime being actively reproductive. Societies which decide to limit the growth of their populations and keep family size down to one or two children will be seeking new roles for women to play. Certainly few societies can afford to keep women in relatively unproductive roles such as housekeepers for 30 or 40 post-family years simply because they were mothers of small children in their twenties. New roles for women pose numerous problems and questions for which science really offers very little by way of answers. For the first time in human history there is a need to know if biologically based sex differences in behavior and psychology exist and if so, whether any of these differences are really biologically fixed or open to the influences of experience and social values.

The increase in life expectancy has placed special burdens on individuals paired in a marriage relationship. In time past many marriages lasted a lifetime, but most lifetimes were much shorter. Only 100 years ago in western societies widows and widowers left with young children were common and many children were raised by stepparents. Today there has been a sharp increase in marriages which end in divorce after children are grown. Perhaps it is better to see these divorces not as symptoms of instability and the breakdown of society but rather as recognition of the fact that at age 40 men and women have whole lifetimes ahead of them and choices made perhaps two decades earlier may no longer by satisfying. Certainly in an evolutionary perspective a relationship which lasted 20 years can hardly be seen as unstable. In other words, a social institution such as marriage, which expects that two people maintain a mutually satisfying relationship for 50 or 60 years, may be asking too much of the emotional capacities of many human beings who are, after all, adapted to much shorter life spans.

Perhaps one of the most important contrasts between small-scale and modern societies is in the size of the group with which the individual has warm, enduring social relations. In time past and in most traditional societies the extended family played a major role in the day-to-day emotional life of the individual. The shape of the extended family varied from society to society but, besides a husband and wife and their young children, it included some of the grandparents, aunts, uncles, cousins, and grandchildren of the man and wife. Often the extended family lived in the same neighborhood or village, if not in the same house or compound. This kin group provided the individual with a group of people, relatively stable in number, familiar, and usually supportive. Interactions outside the kin group were still with neighbors and the amount of daily contact with strangers was minimal. The demands of modern society for economically independent and geographically mobile workers has cut down the extended family until many people are living in the minimal unit, the nuclear family, which includes a husband and wife and their subadult children. This means that emotional life has been focused on a very small number and, if something goes wrong there, it is devastating to the individuals involved. Much of what is thought of as the stress of modern urban life relates to this loss of security and multiplicity of emotional ties. Daily interactions with strangers, frequent turnover in the names, faces, and personalities important in everyday interactions, and stable emotional relations cut down to the nuclear family with only ritual visiting within the kin group—all add to the potential anxiety and stress on the individual.

The cutting down of the family to a minimal size has raised problems in modern societies when divorce, death, or illness removes one of the spouses from a family with small children. In the traditional extended family such misfortune was only a personal loss to the individuals involved. In most situations there were other kin to step in and play the roles of mother or father. Even in everyday matters there were lots of substitutes in emotional life. If mother was too busy, there was always grandmother, aunts, or older sisters, all ready to play the role of mother for a hurt or frightened child. A child raised in the traditional extended family may never in his entire childhood be left with a stranger or expected to sleep alone in a dark or strange room.

Anthropologists have made a number of cross-cultural studies on the ways in which different societies raise children. They have documented a wide variety of childrearing beliefs and techniques ranging from extremes in authoritarianism and permissiveness, to societies where children are thought of as miniature adults and must work, or others where they live in a protected world of play until well past puberty. Healthy, normal children are developed in a variety of contexts according to a number of techniques and beliefs. Nevertheless, there are some aspects of the lives of infants and children that in time past have not varied, and it is only very recently in modern situations where departures from these patterns occur.

First of all, because of the nature of lactation, in time past babies could not be left in a crib and given a bottle propped on a pillow. They were given a minimal amount of daily body contact during feeding at least, and usually they were given much more than that. In gatherer-hunter societies and many traditional societies today infants are carried on their mother's back and sleep with their mothers at night. Furthermore, babies were cared for by their mothers or maternal substitutes and never by strangers. In terms of emotional development there is a vast difference between being cared for in a familiar environment by a group of familiar people such as mother, her sister, father, and grandmother, and being cared for by strangers in a day-care center or orphanage home. Infants need emotional security and they get this before their first year is passed by learning that there are certain individuals, some of whom are always present and ready to play the maternal role. The maternal role is a demanding one. It can be shared between several individuals but the role itself demands the same commitment to the child that the child gives in return—the formation of an emotional bond. An eight-to-ten-hour day in a social environment which is crowded and constantly changing, such as a day-care center for infants, may produce children who cannot form social bonds because they were never given the chance.

It should be pointed out that the success of day-care centers in Israel and China may be due to the fact that the personnel are recruited from the same neighborhood or village as the children and that there is virtually no staff turnover. As the work of Bowlby and others (Bowlby 1969, 1973) suggests, the ability of human beings to bond and form attachments with each other is established during the first year or so of life and emotional privations during the first year cannot be made up later on.

Modern societies such as the United States, Western Europe, Russia, China, Japan, and others are all moving into an era of social experimentation where new answers are being sought to questions never before posed. It is clear that the world has changed; the growth of science and technology has altered social reality so that social systems successful in the past cannot provide solutions to the modern problems of overpopulation, pollution of the environment, nuclear war, and the collapse of world food supplies. Nevertheless, although the physical and social reality of the world has changed radically in the past century, human beings are still virtually the same, equipped with the mind and emotions of the gatherer-hunter. Solutions to modern problems will only work as long as they take into account both realities— the rapidly changing nature of the physical and social world and the more slowly adapting nature of the human species.

TOPICS FOR DISCUSSION OR PROJECTS

1. Try to think of other ways in which the emotional behavior and capacities of human beings no longer match the modern situation. Think over many of the problems of modern societies such as genocide, oppression of minorities, the increase of violent crime, sense of isolation of the individual, obesity, and heart disease and see if you can seek evolutionary discontinuities which may cause or contribute to these problems.

2. What is the future of the family? What new functions might it serve? Do you think other forms of marriage may develop which have no procreative function or division of labor? What functions would they serve?

3. Is there going to be an adaptive value for the division of labor and sex roles in the future? Should childern be encouraged to learn sex roles?

Glossary

Adapt: To come to possess a genetic and/or behavioral system suitable for existing ecological conditions.

Adaptive radiation: Evolutionary divergence of members of a single line into a series of rather different niches or adaptive zones.

Agonistic (adj.): In animal behavior studies refers to behavior patterns used in self-defense and in fighting; the expression of both dominance and submission.

Ape: The common name of the family Pongidae. The human (hominid) line was derived from the Pongid family.

Arboreal: Spends most of time in the trees.

Attention structure: In animal studies, the social organization of a group established by observation of who pays attention to whom and why.

Auditory: Referring to the special sense of hearing.

Australopithecines: A group of early human fossils dated from about 5 to 1 million years ago. The group is characterized by small, ape-sized brains, nearly human bipedalism, the manufacture of stone tools, and hunting.

Bipedalism: Moving about on two legs. In regards to humans this term refers to the special changes in bone and muscle of the lower limbs and pelvis which allow for bipedal striding.

Brachiation: Arm swinging; a way of moving and hanging by the arms involving special anatomical changes at the shoulder, elbow, and wrist joints of the modern apes. It was probably originally a feeding adaptation.

Cerebral cortex: The outer layer of nerve cells of the brain, much expanded among primates and especially humans by the presence of folds or convolutions.

Cerebral dominance: In most animals the left side of the brain is a mirror image of the right side. However, in humans certain specialized functions are represented on only one side (language, skilled hand movements).

Diurnal: Mainly active during the day. A characteristic of monkeys, apes, and humans.

Display: Special behavior pattern which has at least partially evolved to communicate emotional states between individuals.

Dominance: Often defined as the priority of access to some kind of environmental incentive such as food or a good seat.

Environment: The total surroundings of an individual.

Estrus (n.), Estrous (adj.): The period of maximum female sexual receptivity, coinciding with maximum fertility. Although lacking in modern humans, it is a cyclic phenomenon among most mammals. Estrus may occur many times a year in some species, but only every few years in others.

Family: A taxonomic term referring to a grouping between a genus and an order. Members of a family usually share a major pattern of life involving both behavioral and anatomical adaptations.

Fossil: A remaining part of any ancient form of life. Usually it is a hard part which has become mineralized, but prints or casts, when preserved, are also called fossils.

Genes: The unit of the material of inheritance.

Gene pool: The totality of genes of a given population existing at a given time.

Genus: A taxonomic group more inclusive than a species and less than a family. The species of a genus share so many characteristics that they are likely to be in sharp competition if they live in the same area.

Grooming: Among primates, the process of cleaning and picking through the fur and skin. Social grooming is a very important mechanism for maintaining social bonds among monkeys and apes.

Hominid: Taxonomic word meaning belonging to the human family; human.

Homo erectus: An extinct genus of fossil humans living from about 1 million until about 500,000 years ago or later. Considered to be directly ancestral to modern man, *Homo sapiens*.

Knuckle walking: A form of quadrupedal locomotion characteristic of the African apes in which the weight is borne on the flexed knuckles instead of the palm or fingers.

Lateralization: Refers to the establishment of cerebral dominance.

Lesion: A wound, injury, or mass of diseased tissue.

Limbic system: A part of the brain which integrates behavior patterns concerned with vital life functions.

Macaque: The common name for a widely spread genus of Old World monkeys. One species, the rhesus macaque of North India, is the common laboratory monkey in the United States.

Matrifocal group: A group focused on a mother. In primate studies it refers to subgroups composed of individuals related through females.

Monkey: Common name for two separate groups of primates, the Old World and New World monkeys. Refers to an adaptive pattern of behavior based on special senses very similar to man's, a high degree of the development of handiness, and usually life in social groups.

Natal group: The social group into which an individual is born.

Natural selection: The principal mechanism of evolutionary change whereby those individuals best adapted to the environment contribute more offspring to the following generation than do others so that, if their characteristics are inheritable, the composition of the gene pool will change.

Niche (ecological):   The totality of environmental factors into which a species fits; the specific way an organism utilizes its environment.

Nocturnal: Mainly active at night. Among primates it is typical of many of the more primitive species.

Olfactory: Refers to the sense of smell.

Phylogenetic: Refers to the evolutionary history of a taxonomic group.

Pleistocene: A geological time period which witnessed the evolution of humans.

Primate: The order of mammals to which humans belong. The order is characterized by a fundamental adaptation of climbing by grasping. "Higher primate" refers to the Old World monkeys, apes, and humans, a grouping of families which hold many characteristics in common such as similar development of the special senses.

Prosimian: The earliest and most primitive forms of primates.

Quadrupedal: Moving about on four legs. Among the primates it is typical of primitive forms and monkeys.

Sensory modality: Anyone of the various special senses: vision, hearing, touch, or smell.

Sexual dimorphism: The characteristic difference between the sexes of a single species.

Sibling: Brother or sister.

Specialization: In the context of evolutionary studies, a character evolved for a particular and limited function—the opposite of a generalized character.

Speciation: The evolutionary process by which a single species splits up into more than one species, usually through geographical and ecological separation.

Species: The basic taxon among sexually reproducing animals. The lines of genetic communication are open within a species, and it is more or less genetically isolated from other closely related species through geographic or social barriers.

Tactile: Referring to the special sense of touch.

Taxonomy: The science of systematic classification of living things in such a way as to indicate their evolutionary relationship to each other.

Terrestrial: Living on the ground.

Territorialism: The maintaining of exclusive access to a home range from other members of the species who do not belong to the same social group, usually through active defense or aggressive display.

Bibliography

Andrew, R. J., 1962, "The situations that evoke vocalization in primates," *Annals of the New York Academy of Science*, 102:296–315.

Bernstein, I. S., 1970, "Primate status hierarchies," in *Primate Behavior*, L. A. Rosenblum, ed., Volume 1, pp. 71–109. New York: Academic Press.

Bowlby, John, 1969, 1973, *Attachment and Loss*, Volumes I and II. New York: Basic Books.

Brown, J., 1970, "A note on the division of labor by sex," *American Anthropologist*, 72:1073–1078.

Cartmill, M., 1974, "Rethinking primate origins," *Science*, 184:436–443.

Chance, M. R. A., 1967, "Attention structure as the basis of primate rank orders," *Man* 2:503–518.

Chance, M. R. A., and C. J. Jolly, 1970, *Social Groups of Monkeys, Apes and Men*. New York: Dutton.

DeVore, I., and K. R. L. Hall, 1965, "Baboon ecology," in *Primate Behavior: Field Studies of Monkeys and Apes*, I. DeVore, ed., pp. 20–52. New York: Holt, Rinehart and Winston, Inc.

Dolhinow, P. J., and N. Bishop, 1970, "The development of motor skills and social relationships among primates through play," in *Minnesota Symposia on Child Psychology*, J. P. Hill, ed., 4:141–198.

Eibl-Eibesfeldt, I., 1975, *Ethology: The biology of behavior*, 2d ed. New York: Holt, Rinehart and Winston, Inc.

Eisenberg, J. F., N. A. Muckenhirn, R. Rudran, 1972, "The relation between ecology and social structure in primates," *Science*, 176:863–874.

Ellefson, J. O., 1968, "Territorial behavior in the common white-handed gibbon, *Hylobates lar Linn*," in *Primates, Studies in Adaptation and Variability*, P. Jay, ed., pp. 180–199. New York: Holt, Rinehart and Winston, Inc.

Gardner, B. T., and R. A. Gardner, 1971, "Two-way communication with an infant chimpanzee," in *Behavior of Nonhuman Primates*, A. M. Schrier and F. Stollnitz, eds., Volume 4, pp. 117–185. New York: Academic Press.

Gartlan, J. S., and C. K. Brain, 1968, "Ecology and social variability in *Cercopithecus aethiops*," in *Primates: Studies in Adaptation and Variability*, P. Jay, ed., pp. 253–292. New York: Holt, Rinehart and Winston, Inc.

Geschwind, N., 1972, "Language and the brain," *Scientific American*, 794:76–83.

Hall, K. R. L., 1963, "Observational learning in monkeys and apes," *British Journal of Psychology*, 54:201–226.

———, 1968, "Social learning in monkeys," in *Primates: Studies in Adaptation and Variability*, P. Jay, ed., pp. 383–397. New York: Holt, Rinehart and Winston, Inc.

Hall, K. R. L., and I. DeVore, 1965, "Baboon social behavior," in *Primate Behavior: Field Studies of Monkeys and Apes*, I. DeVore, ed., pp. 53–110. New York: Holt, Rinehart and Winston, Inc.

Hamburg, D., 1962, "Relevance of recent evolutionary changes to human stress biology," in *Social Life of Early Man*, S. L. Washburn, ed., pp. 278–288. Chicago: Aldine.

———, 1963, "Emotions in the perspective of human evolution," in *Expression of the Emotions in Man*, P. Knapp, ed., pp. 300–317. New York: International Universities.

Harding, R. S. O., 1973, "Predation by a troop of olive baboons (*Papio anubis*)," *American Journal of Physical Anthropology*, 38:587–591.

Hinde, R. A., and T. E. Rowell, 1962, "Communication by postures and facial expressions in the rhesus monkey (Macaca mulatta)," *Proceedings of the Zoological Society of London*, 138:1–21.

Isaac, Glynn, 1968, "Traces of Pleistocene hunters," in *Man the Hunter*, R. B. Lee and I. DeVore, eds., pp. 253–261. Chicago: Aldine.

———, 1971, "The diet of early man: Aspects of archaeological evidence from lower and middle Pleistocene sites in Africa," *World Archaeology*, 2:278–299.

Itani, J., 1958, "On the acquisition and propagation of a new food habit in the troop of Japanese monkeys at Takasakiyama," *Primates*, 1:84–98.

Jolly, A., 1972, *The Evolution of Primate Behavior*. New York: Macmillan.

Kaufman, I. C., 1973, "The role of ontogeny in the establishment of species-specific patterns," *Early Development*, 51:381–397.

Kaufman, I. C., and L. A. Rosenblum, 1969, "Effects of separation from mother on the emotional behavior of infant monkeys," *Annals of the New York Academy of Science*, 159:681–695.

Kawai, M., 1965, "Newly acquired pre-cultural behavior of the natural troop of Japanese monkeys on Koshima islet," *Primates*, 6:1–30.

Kawamura, S., 1958, "Matriarchial social ranks in the Minoo-B troop: A study of the social rank system of Japanese monkeys," *Primates*, 1:149–156.

———, 1959, "The process of sub-culture propagation among Japanese macaques," *Primates*, 2:43–55.

Koford, C., 1963, "Rank of mothers and sons in bands of rhesus monkeys," *Science*, 141:356–357.

———, 1966, "Population changes in rhesus monkeys: Cayo Santiago, 1960–1964," *Tulane Studies in Zoology*, 13:1–7.

Koyama, N., 1970, "Changes in dominance rank and division of a wild Japanese monkey troop in Arashiyama," *Primates*, 11:335–391.

Kruuk, H., 1972, *The Spotted Hyena*. Chicago: University of Chicago Press.

Kummer, H., 1968, "Two variations in the social organization of baboons," in *Primates: Studies in Adaptation and Variability*, P. Jay, ed., pp. 293–312. New York: Holt, Rinehart and Winston, Inc.

———, 1971, *Primate Societies*. Chicago: Aldine.

Lancaster, J. B., 1968a, "On the evolution of tool-using behavior," *American Anthropologist*, 70:56–66.

———, 1968b, "Primate communication systems and the emergence of human language," in *Primates: Studies in Adaptation and Variability*, P. Jay, ed., pp. 439–457. New York: Holt, Rinehart and Winston, Inc.

———, 1971, "Play-mothering: The relations between juvenile females and young infants among free-ranging vervet monkeys (*Cercopithecus aethiops*)," *Folia Primatologia*, 15:161–182.

Lawick-Goodall, J. van, 1968, "A preliminary report on expressive movements and communication in the Gombe Stream chimpanzees," in *Primates: Studies in Adaptation and Variability*, P. Jay, ed., pp. 313–374. New York: Holt, Rinehart and Winston, Inc.

———, 1970, "Tool-using in primates and other vertebrates," in *Advances in the Study of Behavior*, D. S. Lehrman, R. A. Hinde, and E. Shaw, eds., Volume 3, pp. 195–249. New York: Academic Press.

———, 1971, *In the Shadow of Man*. New York: Houghton Mifflin.

Lee, R. B., and I. DeVore, 1968, "Problems in the study of hunters and gatherers," in *Man the Hunter*, R. B. Lee and I. DeVore, eds., pp. 3–13. Chicago: Aldine.

Le Gros Clark, W. E., 1967, *Man-apes or Ape-men?: The story of discoveries in Africa*. New York: Holt, Rinehart and Winston, Inc.

Lenneberg, Eric, 1967, *Biological Foundations of Language*. New York: Wiley.

Lieberman, P., 1968, "Primate vocalizations and human linguistic ability," *Journal of the Acoustical Society of America*, 44:1574–1584.

McGrew, W. C., and C. Tutin, 1973, "Chimpanzee tool use in dental grooming," *Nature*, 241:477–478.

MacLean, P., 1954, "Studies on limbic system ("Visceral Brain") and their bearing on psychosomatic problems," in *Recent Developments in Psychosomatic Medicine*, E. Wittkower and R. Cleghorn, eds., pp. 101–125. London: Pitman.

Marler, P., 1965, "Communication in monkeys and apes," in *Primate Behavior: Field*

Studies of Monkeys and Apes, I. DeVore, ed., pp. 544–584. New York: Holt, Rinehart and Winston, Inc.

————, 1969, "Vocalizations of wild chimpanzees," in *Proceedings of the Second International Congress of Primatology*, C. R. Carpenter, ed., 1:94–100. Basel: S. Karger.

Mason, W. A., 1960, "The effects of social restriction on the behavior of rhesus monkeys, I. Free social behavior," *Journal of Comparative and Physiological Psychology*, 53:582–589.

————, 1961a, "II. Tests of gregariousness," *Journal of Comparative and Physiological Psychology*, 54:287–290.

————, 1961b, "II. Dominance tests," *Journal of Comparative and Physiological Psychology*, 54:694–699.

Missakian, E. A., 1972, "Genealogical and cross-genealogical dominance relations in a group of free-ranging rhesus monkeys (*Macaca mulatta*) on Cayo Santiago," *Primates*, 13:169–181.

Neville, M. K., 1968, "A free-ranging rhesus monkey troop lacking adult males," *Journal of Mammalogy*, 49:771–773.

Paterson, J. D., 1973, "Ecologically differentiated patterns of aggressive and sexual behavior in two troops of Ugandan baboons, *Papio anubis*," *American Journal of Physical Anthropology*, 38:641–648.

Pilbeam, David, 1972, *The Ascent of Man: An Introduction to Human Evolution*. New York: MacMillan.

Premack, D., 1971, "On the assessment of language competence in the chimpanzee," in *Behavior of Nonhuman Primates*, A. M. Schrier and F. Stollnitz, eds., vol. 4, pp. 186–228. New York: Academic Press.

Rosenblum, L. A., I. C. Kaufman, and A. J. Stynes, 1964, "Individual distance in two species of macaque," *Animal Behavior*, 12:338–342.

Rowell, T. E., 1962, "Agonistic noises of the rhesus monkey (*Macaca mulatta*)", *Symposium of the Zoological Society of London*, 8:91–96.

————, 1967, "Variability in the social organization of primates," in *Primate Ethology*, D. Morris, ed., pp. 219–235. Garden City, New York: Doubleday.

————, 1969, "Long-term changes in a population of Ugandan Baboons," *Folia primatologia*, 11:241–254.

————, 1972, *Social Behavior of Monkeys*. Baltimore: Penguin.

————, 1973, "Social organization of wild Talapoin monkeys," *American Journal of Physical Anthropology*, 38:593–598.

————, 1974, "Contrasting adult male roles in different species of nonhuman primates," *Archives of Sexual Behavior*, 3:143–149.

Rowell, T. E., and R. A. Hinde, 1962, "Vocal communication by the rhesus monkey (*Macaca mulatta*)," *Proceedings of the Zoological Society of London*, 138:279–294.

Sade, D. S., 1965, "Some aspects of parent–offspring and sibling relations in a group of rhesus monkeys, with a discussion of grooming," *American Journal of Physical Anthropology*, 23:1–18.

————, 1967, "Determinants of dominance in a group of free-ranging rhesus," in *Social Communication among Primates*, S. A. Altmann, ed., pp. 99–114. Chicago: University of Chicago Press.

Schaller, G., 1963, *The Mountain Gorilla*. Chicago: University of Chicago Press.

Schaller, G., and G. Lowther, 1969, "The relevance of carnivore behavior to the study of early hominids," *Southwestern Journal of Anthropology*, 25:307–341.

Schultz, A. H., 1969, *The Life of Primates*, London: Weidenfeld and Nicholson.

Simons, E. L., 1972, *Primate Evolution: An introduction to man's place in nature*. New York: Macmillan.

Singh, S. D., 1969, "Urban monkeys," *Scientific American*, 221:108–116.

Skinner, B. F., 1966, "The phylogeny and ontogeny of behavior," *Science*, 153:1205–1213.

Struhsaker, T. T., 1967, "Auditory communication among vervet monkeys (*Cercopithecus aethiops*)," in *Social Communication Among Primates*, S. A. Altmann, ed., pp. 281–324. Chicago: University of Chicago Press.

Struhsaker, T. T., and J. S. Gartlan, 1970, "Observations on the behavior and ecology of the patas monkey (*Erythrocebus patas*) in the Waza Reserve, Cameroons," *Journal of Zoology* (London), 161:49–63.

Struhsaker, T. T., and P. Hunkeler, 1971, "Evidence of tool-using by chimpanzees in the Ivory Coast," *Folia primatologia*, 15:212–219.

Teleki, G., 1973, *The Predatory Behavior of Wild Chimpanzees*. Lewisburg, Pennsylvania: Bucknell University Press.

Thorpe, W. H., 1963, *Learning and Instinct in Animals*, 2d ed. London: Methuen.

Tobias, P. V., 1963, "Cranial Capacity of *Zinjanthropus* and other Australopithecines," *Nature*, 197:743–746.

———, 1971, *The Brain in Hominid Evolution*. New York: Columbia University Press.

Washburn, S. L., 1960, "Tools and human evolution," *Scientific American*, 203:3–15.

———, 1968, *The Study of Human Evolution*, Condon Lectures, Oregon State System of Higher Education. Eugene: University of Oregon Press.

Washburn, S. L., and C. S. Lancaster, 1968, "The evolution of hunting," in *Man the Hunter*, R. B. Lee and I. DeVore, eds., pp. 293–303. Chicago: Aldine.

Washburn, S. L., and R. Moore, 1974, *Ape into Man: A Study of Human Evolution*. Boston: Little, Brown.

Winter, P., D. Ploog, and J. Latta, 1966, "Vocal repertoire of the squirrel monkey (*Saimiri sciureus*), its analysis and significance," *Experimental Brain Research*, 1:359–384.

Recommended Readings

The following list of books were selected for being readable and up-to-date, and most are available in paperback editions. They represent good quality research and analysis and are written for college students and general readers.

Evolution
Dobzhansky, T., 1973, *Genetic Diversity and Human Equality*. New York: Basic Books.
Stebbins, G. L., 1966, *Processes of Organic Evolution*. Englewood Cliffs, New Jersey: Prentice-Hall.

Primate and Human Evolution
Jolly, A., 1972, *The Evolution of Primate Behavior*. New York: Macmillan.
Simons, E. L., 1972, *Primate Evolution: An Introduction to Man's Place in Nature*. New York: Macmillan.
Washburn, S. L., and R. Moore, 1974, *Ape into Man: A Study of Human Evolution*. Boston: Little, Brown.

Social Behavior of Monkeys, Apes, and Man
Alland, A. Jr., 1972, *The Human Imperative*. New York: Columbia University Press.
Altmann, S. A. (ed.), 1967, *Social Communication among Primates*. Chicago: University of Chicago Press.
Harlow, H. F., 1971, *Learning To Love*. New York: Ballantine.
Kummer, H., 1971, *Primate Societies*. Chicago: Aldine.
Lawick-Goodall, J. van, 1971, *In the Shadow of Man*. New York: Dell.
Montagu, A. (ed.), 1973, *Man and Aggression*, 2d ed. New York: Oxford University Press.
Poirier, F. E., 1972, *Primate Socialization*. New York: Random House.
Rowell, T., 1972, *Social Behavior of Monkeys*. Baltimore: Penguin.
Southwick, C. H. (ed.), 1970, *Animal Aggression: Selected Readings*. New York: Van Nostrand.
Wickler, W., 1972, *The Sexual Code*. Garden City, New York: Doubleday.

Relevant Case Studies[*]

The Case Studies listed below provide good ethnographic data on human behavior patterns in small groups. They can be used to provide material for papers or discussions on topics such as gathering and hunting subsistence patterns, life in small tribal groups, war and violence, cooperation, territoriality, socialization, and the roles of males and females.

Case Studies in Cultural Anthropology

Chagnon, Napoleon A., 1966, *Yanomamö: The Fierce People.* This is an ethnographic account of a South American tribal group which lives in small villages in tropical forests. Subsistence is based on slash-and-burn horticulture and hunting. The Yanomamö represent an extreme example of warfare, raiding, and the expression of aggression in a small-scale society. There is a good section on the division of labor between males and females and adults and young in a social system in which war, raids, abduction, and violence are facts of daily life.

Dentan, Robert K., 1968, *The Semai: A Nonviolent People of Malaya.* This Case Study includes an excellent description of social and marital relations, sex, and aggression among a group of Malaysian slash-and-burn horticulturalists known for their nonviolent ways of coping with social problems.

Hart, C. W. M., and Arnold R. Pilling, 1960, *The Tiwi of North Australia.* This Case Study is an account of the life of a group of Australian aborigines living on an island near Australia. The Tiwi are gatherer-hunters. There are especially interesting descriptions of marriage and relationships between the sexes in a society in which women represent the primary social valuables for which men compete.

Keiser, R. L., 1969, *The Vice Lords: Warriors of the Streets.* This study describes modern street society in an industrialized urban center. Small group social relations based on male dominance are well described. The study forms an interesting comparison with the Yanomamö, another group in which the expression of aggression to dominate others is highly rewarded.

Case Studies in Education and Culture.

Leis, Philip, 1972, *Enculturation and Socialization in an Ijaw Village.* This book is about growing up in a small tribal village in the Niger Delta. It includes descriptions of social development of the young child, the close relationship between mother and infant, the role played by child nurses, play groups, responsibility for household tasks, puberty, and the assumption of adult roles.

Basic Anthropology Units

Friedl, Ernestine, 1975, *Women and Men: An Anthropologist's View.*

Kinsey, W. *Invitation to Physical Anthropology.* In preparation.

* Edited by George and Louise Spindler, and published by Holt, Rinehart and Winston, Inc., as Case Studies in Cultural Anthropology, Case Studies in Education and Culture, and Basic Anthropology Units.

Recommended Films
on Primate Social Behavior

Monkeys, Apes and Man by Jeff Myrow. 1971. 16 mm, color, optical sound, 52 minutes. Films Inc., 1144 Wilmette Ave., Wilmette, Ill. 60091.
 Excellent general introduction to primate behavior studies. Includes footage on predation by chimpanzees, a chimpanzee attack on a stuffed leopard, chimpanzee tool-using, the role of predation pressure in shaping baboon society, dominance in baboons.

The Baboon Troop by I. DeVore, 1966. 16 mm, color, sound, 23 minutes.
Dynamics of Male Dominance in a Baboon Troop by I. DeVore, 1966. 16 mm, color, sound, 30 minutes.
 The above two films are available from Modern Learning Aids, 1212 Avenue of the Americas, New York, N.Y. 10036.
Evening Activity by I. DeVore, 1966. 16 mm, color, sound, 6 minutes.
Animals in Amboseli by I. DeVore, 1966. 16mm, color, sound, 20 minutes.
Younger Infant: Birth to Four Months by I. DeVore, 1966. 16 mm, color, sound, 10 minutes.
Older Infant: Four Months to One Year by I. DeVore, 1966. 16 mm, color, sound, 8 minutes.
 The above films are available from Audio-Visual Services, Pennsylvania State University, University Park, Pa. 16802.

The six films on baboons by DeVore are recommended for their general coverage of a monkey social system. The photography is excellent and the sound in the films on infants is dramatically real and immediate. The first two films are excellent for general coverage on baboon social behavior and for a careful analysis of the complexities of a dominance system. The latter four films are interesting in that they give a good impression of field experience. There is a minimal amount of editing of the original footage so that the student gets the impression of what it is really like to observe natural behavior of free-ranging animals.

Behavior and Ecology of Vervet Monkeys by T. T. Struhsaker, 1971. 16 mm, color, sound, 40 minutes. Rockefeller University Film Service, 241 West 17th Street, New York, N.Y. 10011.

 This film is a good, balanced presentation of vervet monkey life. It is useful in contrast to the films on baboon life because vervets share the baboon's habitat but follow different social adaptations due to their separate phylogenetic history. There is some good footage on dominance relations, infant development and caretaking by juvenile females, and territorial defense by the entire group.

Miss Goodall and the Wild Chimpanzees by Jane van Lawick-Goodall, 1968. 16 mm, color, sound, 28 minutes. Audio-Visual Services, Pennsylvania State University, University Park, Pa. 16802.
Chimpanzees of the Gombe National Park by Jane van Lawick-Goodall, 1971. 16 mm, color, sound, 10 minutes.

A Chimpanzee Family by Jane van Lawick-Goodall, 1971. 16 mm, color, sound, 7 minutes. Above two films are available from Educational Development Center, Film Distribution Center, 55 Chapel Street, Newton, Mass. 02160.

These three films give an excellent general coverage of the behavior of wild chimpanzees. Footage includes scenes of tool using, aggressive interactions, facial expressions and gestures, and mother–infant and mother–offspring and sibling relationships. The parallels with human nonverbal communication and family interactions are particularly striking.